W9-CHE-961

Effects of Conservation Tillage on Groundwater Quality

Nitrates and Pesticides

Terry J. Logan
James M. Davidson
James L. Baker
Michael R. Overcash

Library of Congress Cataloging-in-Publication Data

Effects of conservation tillage on groundwater quality.

Includes bibliographies and index.
1. Conservation tillage—Environmental aspects.
2. Water, Underground—Quality. 3. Nitrates—
Environmental aspects. 4. Pesticides—Environmental
aspects. I. Logan, Terry J.
S604.E35 1987 631.5′1 87-3716
ISBN 0-87371-080-0

LEWIS PUBLISHERS, INC.
121 South Main Street, P.O. Drawer 519, Chelsea, Michigan 48118

PRINTED IN THE UNITED STATES OF AMERICA

Preface

During the last decade, there has been a major shift by U.S. farmers away from inversion tillage, such as moldboard plowing, towards systems with reduced tillage, various versions of which are collectively termed conservation tillage (CT). CT has been defined by the U.S. Soil Conservation Service as providing 30% or more crop residue cover of the land surface at the time of planting, and the most recent statistics indicate that about 30% of cropland in the continental United States is planted with some form of conservation tillage. These practices have been most widely adopted by grain farmers in the Corn Belt and are most often associated with corn, soybean, and wheat production.

Tillage is most commonly performed to prepare a suitable seedbed for germination, to bury excessive residues, to control weeds, and to incorporate agricultural chemicals. The successful substitution of conservation tillage for conventional tillage requires, among other factors, that planting equipment can place the seed in a rough-surface soil partially or completely covered with residues, that weeds and pests can be controlled by chemical means, and that fertilizer and other agricultural chemicals can be applied and be effective with reduced or no tillage incorporation. The significant expansion of conservation tillage acreage in the United States during the last 10 years is testimony to the successful development of equipment, chemicals, and management to meet these requirements. Major equipment manufacturers today offer a full line of no-till and/or conservation tillage planters, chisel plows, ridge planters, or other implements that permit consistent seed placement and high subsequent germination in a wide range of rough seedbed conditions. Nonselective "knock-down" herbicides, as well as postemergent grass and broadleaf weed control chemicals, are available today to provide critical weed control needed in conservation tillage systems. Injection equipment for the placement of fertilizer and pesticides in soil beneath crop residues is also increasingly available. These developments have greatly increased the probability of achieving crop yields comparable to those obtained with conventional clean-tillage methods, and have raised the confidence of farmers in their suitability for modern high-yield crop production. This increased acceptability by farmers, together with the real and perceived savings in time, labor, energy, and other production inputs, can explain the sustained growth in adoption of conservation tillage practices by U.S. grain farmers.

Protection of the soil surface by residue cover from raindrop impact and runoff is highly effective in reducing soil loss and sediment loads. The very high surface residue coverage provided by no-till can reduce water erosion by as much as 90%, and residue cover in the 30% range by which conservation tillage is defined will usually reduce soil loss by more than 50%. These decreases in erosion and in sediment loads from agricultural land are accompanied by reductions in sediment-bound nonpoint-source pollutants, of which phosphorus and pesticides are the most prominent. The ability to achieve significant reductions in these important pollutants by adoption of farmer-acceptable practices has led to the promotion of conservation tillage as a "best management practice" for nonpoint-source pollution control. Federal and state conservation, environmental control, and agricultural agencies have been unanimous in their promotion of conservation tillage as a practice that can benefit the farmer, conserve the soil resource, and protect the environment.

In the last several years, concern has been raised regarding the environmental soundness of conservation tillage systems. Three issues most commonly raised are these: Does residue cover in conservation tillage reduce surface runoff and increase infiltration, with a resulting potential increase in groundwater contamination by soluble agricultural chemicals, chiefly nitrate and some pesticides? Does elimination of tillage for weed control increase use of herbicides that may leach to groundwater? Does increased residue cover provide a refuge for insects which will increase insecticide use? Countering these arguments are claims that reduced runoff occurs only on more permeable soils and may even increase with conservation tillage on fine-textured soils with low hydraulic conductivities; that the rates and kinds of herbicides used in conservation tillage are no different than those used with clean tillage; and that pest control can be achieved with integrated pest management programs and crop rotations without an increase in insecticide use.

Research has been conducted on conservation tillage in the United States since the early 1960s, and various systems have been used by farmers since the late 1960s. Knowledge gained from research and from on-farm use in the last 20 years has led to many of the assertions made about the potential environmental effects of conservation tillage practices. Unfortunately, there has been a tendency to generalize regarding the environmental effects over the wide range of soil, climate, crop, and tillage management conditions found in the "real world."

Many of the effects on the environment attributed to conservation tillage are based on research which compared the two extremes of the tillage continuum—clean inversion tillage and no-till. For the purposes of identifying and quantifying effects, the choice of extreme contrasts by the researcher is logical. However, conservation tillage as practiced by the majority of farmers involves much less residue cover than no-till—perhaps closer to the minimum of 30% cover by which conservation tillage is defined—and when no-till is used, some tillage during the growing season in the form of cultivation is often performed. The findings of the workshop indicate that some degree of tillage can greatly reduce the differences between no-till and inversion tillage of soil properties and processes that affect environmental quality. It is, therefore, inaccurate to extrapolate the results of research on no-till to situations where the majority of conservation tillage practices involve some kind of soil disturbance and residue incorporation.

This book focuses on the potential contamination of groundwater, and to a lesser extent surface water, by nitrate and pesticides as a result of widespread shifts from inversion tillage to conservation tillage. Concern for groundwater contamination by agricultural and industrial chemicals commands national attention today. Among these, nitrate and pesticides receive considerable attention because of the large quantities used by farmers and because of the high mobilities of nitrate and some of the pesticides in soil. The question posed here is not whether these compounds migrate to groundwater as a result of agricultural practices, or the extent of groundwater contamination. Rather, the authors were asked to consider whether extensive shifts from inversion tillage to conservation tillage would increase the potential for groundwater contamination by nitrate and pesticides. These authors were selected for their experience with conservation tillage systems.

The first three chapters provide background on the extent of conservation tillage practices as currently used in the United States, and pesticide and nitrogen fertilizer use with conservation tillage. These are followed by a review chapter on nonpoint-source pollution in the Great Lakes, with particular emphasis on nitrogen and pesticides. A series of lead and discussion chapters are then presented on effects of conservation tillage on physical, chemical, and biological processes in soil and on surface and groundwater hydrology. The next lead and discussion chapters concern nitrogen and pesticide fate and transformations in soil, and the final chapter is on interactions of conservation tillage and agricultural waste management.

Conservation tillage use is concentrated in the Corn Belt region of the north central United States and the Great Lakes states, and this practice has been promoted as a means of reducing nonpoint phosphorus source loads to the Great Lakes. The impacts of large-scale implementation of this practice on groundwater quality in this region was a major stimulus for this book, and much of the discussion is centered on research conducted in this general geographical area. An attempt has been made, however, to include representatives from other areas, such as the Southeast, Northeast, and Great Plains, to provide a broader perspective to this issue and to determine, where possible, major differences among geographical regions in factors affecting conservation tillage and major soil processes.

T. J. Logan
J. M. Davidson
J. L. Baker
M. Overcash

Acknowledgments

The editors are indebted to the U.S. Environmental Protection Agency, Great Lakes National Program Office (Chicago, IL), for technical and grant support. Ralph Christensen, Kent Fuller, and Lynn Shuyler were specifically instrumental during the conceptual and planning period of our workshop. Their understanding of the importance of the topic of the workshop and this book is appreciated.

The editors appreciate the assistance Mrs. Tammy Langford provided in coordinating travel for speakers and hotel arrangements for the workshop. Also, the assistance of Mrs. Barbara Kurtz and Miss Judy Kite in arranging for retyping of all chapters in the book and coordinating the final stages of the book is appreciated.

Mrs. Lisa Hurewitz typed the final manuscripts for the book and the editors sincerely thank her for her effort and patience.

The editors wish to thank all speakers and authors of the chapters for their presentations at the workshop and promptness in providing manuscripts of their talks. A special thank-you is also extended to all those who attended the workshop and participated in the discussions.

T. J. Logan
J. M. Davidson
J. L. Baker
M. Overcash

Terry J. Logan is professor of soil chemistry at The Ohio State University. He has a B.S. degree in soil science from California Polytechnic State University, San Luis Obispo, and M.S. and PhD degrees in soil chemistry from The Ohio State University. He teaches soil chemistry, and his research has included the environmental chemistry of phosphorus, diffuse source losses of sediment and nutrients from agricultural land, chemistry and plant availability of nutrients and trace elements in municipal sewage sludge, and soil erosion from steeplands in the humid tropics. He currently serves as editor of the *Journal of Environmental Chemistry.*

James M. Davidson is professor of soil physics and dean for research in the Institute of Food and Agricultural Sciences at the University of Florida. His research activities include measurement and simulation of pesticide sorption and movement in water-saturated and -unsaturated soils, and development and evaluation of mathematical relationships for describing the fate of various nitrogen species associated with commercial fertilizers and pesticides. He received a PhD in soil physics in 1965 from the University of California at Davis, and his M.S. and B.S. degrees from Oregon State University. Dr. Davidson has served as chairman of the Environmental Quality Division of the American Society of Agronomy, and as associate editor for the *Soil Science Society of America Journal.* He also serves on various national and international committees concerned with the fate of water and chemicals in the environment. Dr. Davidson is author of more than 100 scientific papers in his research field. With Michael R. Overcash, he coedited *Environmental Impact of Nonpoint Source Pollution,* published in 1980 by Ann Arbor Science Publishers.

James L. Baker is a professor in the Department of Agricultural Engineering at Iowa State University. His research has been involved with farm management practices and equipment (e.g., a point-injector fertilizer applicator) that result in protection of our soil resource, efficient use of agricultural chemicals and energy resources, and improved water quality of agricultural drainage. He comes from a farm/ranch background with a PhD in physical chemistry from Iowa State University and a B.S. degree in chemistry from the South Dakota School of Mines and Technology. He is author or coauthor of 25 papers concerned with the transport of nutrients and pesticides with surface runoff and subsurface drainage as affected by management practices, particularly conservation tillage and nitrogen fertilization. He also serves on several committees and has made presentations, on both a local and national level, concerned with groundwater contamination from agriculture.

Michael R. Overcash is a professor of chemical engineering and a professor of biological and agricultural engineering at North Carolina State University. He was recently awarded the Distinguished Scientist Recognition from the U.S. Environmental Protection Agency. Dr. Overcash has contributed to the modeling of nonpoint sources from agriculture and the understanding of specific organic compound behavior in plant-soil systems.

Contents

SECTION I.
OVERVIEW OF CONSERVATION TILLAGE SYSTEMS IN THE UNITED STATES

SECTION II.
EFFECT OF CONSERVATION TILLAGE SYSTEMS ON SOIL PHYSICAL, CHEMICAL, AND BIOLOGICAL PROCESSES

SECTION III.
EFFECT OF CONSERVATION TILLAGE SYSTEMS ON FATE AND TRANSPORT OF APPLIED PESTICIDES AND NITROGEN FERTILIZERS

SECTION IV.
SELECTED TOPICS ON CONSERVATION TILLAGE SYSTEMS

List of Figures

List of Tables

Effects of Conservation Tillage on Groundwater Quality

Nitrates and Pesticides

SECTION I

OVERVIEW OF CONSERVATION TILLAGE SYSTEMS IN THE UNITED STATES

CHAPTER 1

OVERVIEW OF CONSERVATION TILLAGE

J. V. Mannering,
Purdue University, West Lafayette, Indiana

D. L. Schertz,
USDA-Soil Conservation Service, Washington, D. C.

B. A. Julian,
Conservation Tillage Information Center,
Ft. Wayne, Indiana

DEFINITION OF CONSERVATION TILLAGE

Confusion exists over the meaning of the term "conservation tillage." In publications and talks the term is used interchangeably with minimum till, reduced till, lo-till, mulch-till, no-till, etc. Then there is the confusion caused by some writers who use only the two terms "no-till" and "conservation tillage". This approach generally defines no-till as narrow strip tillage and lumps everything else into conservation tillage (chisel, till-plant, disk, field cultivator, etc.). This approach simplifies writing and discussion, but is not specific enough as to what conservation tillage really is.

Another complicating factor is that there is somewhat different terminology used by those working in primarily rowcrop growing areas and those in primarily small grain areas. The "row croppers" commonly use the terms no-till, ridge-till, chisel, field cultivators and disks as forms of conservation tillage. The terminology of the small grain people consists of stubble mulch farming which includes various types of disks, chisel plows, mulch treaders, sweep plows or blades, and rodweeders, and other categories which include "ecofallow" and

Effects of Conservation Tillage on Groundwater Quality: Nitrates and Pesticides, Terry J. Logan et al., eds.

"direct drill." Additional discussion of tillage nomenclature can be found in articles by Mannering and Fenster (1983), Allen and Fenster (1986), Siemens, et al. (1985) and in the SCSA Resource Conservation Glossary (1982).

Because of the confusion on terminology, the Conservation Tillage Information Center (CTIC) defined conservation tillage as an umbrella term and further defined types of systems that fit under the umbrella. This approach permitted them to do a national survey on tillage systems which not only included row crop and small grain agriculture but also included forage crops, vegetable and truck crops. This successful annual survey has been underway since 1982 (CTIC, 1982-1985).

Because the CTIC definitions are accepted on a national basis and since national survey data is being collected based on this terminology, it was decided to use these definitions in this workshop. These are given below.

Conservation Tillage

Conservation tillage is defined as any tillage and planting system that maintains at least 30 percent of the soil surface covered by residue after planting to reduce soil erosion by water; or where soil erosion by wind is the primary concern, maintains at least 450 kg (1,000 pounds) per acre of flat small grain residue equivalent on the surface during the critical erosion period.

Types of Conservation Tillage

1. No-till or slot planting: The soil is left undisturbed prior to planting. Planting is completed in a narrow seedbed approximately 2-8 cm wide. Weed control is accomplished primarily with herbicides.

2. Ridge-till (includes no-till on ridges): The soil is left undisturbed prior to planting. Approximately 1/3 of the soil surface is tilled at planting with sweeps or row cleaners. Planting is completed on ridges usually 10-15 cm higher than the row middles. Weed control is accomplished with a combination of herbicides and cultivation. Cultivation is used to rebuild ridges.

3. Strip-till: The soil is left undisturbed prior to planting. Approximately 1/3 of the soil surface is tilled at planting time. Tillage in the row may consist of a rototiller, in-row chisel, row cleaners, etc. Weed control is accomplished with a combination of herbicides and cultivation.

4. Mulch-till: The total surface is disturbed by tillage prior to planting. Tillage tools such as chisels, field

cultivators, disks, sweeps, or blades are used. Weed control is accomplished with a combination of herbicides and cultivation.

5. Reduced-till: Any other tillage and planting system not covered above that meets the 30 percent residue requirement.

Conventional Tillage

This refers to the combined primary and secondary tillage operation normally performed in preparing a seedbed for a given crop grown in a given geographical area. Since operations vary considerably under different climatic, agronomic and other conditions, the definition also varies from one region to another. For example, in the central Corn Belt this might be fall moldboard plowing, followed in the spring by disking and field cultivating prior to planting. While in the Great Plains wheat producing areas, the primary tillage might be a chisel or sweep-type plow after wheat harvest followed by an assortment of other forms of secondary tillage such as disks, rod-weeders, harrows, etc. to prepare the seedbed prior to planting. In the context of this discussion, even though a primary tillage tool other than a moldboard plow is used, there may be insufficient residue on the soil surface after planting to qualify as "conservation tillage" if secondary tillage is used.

Conventional tillage is often used as the "standard" or "check" in experiments to assess the potential of other systems in a given area. It involves a set of operations that prepares a seedbed having essentially no plant residue left on the soil surface. Many conventional systems leave the surface bare, particularly those based on the use of the moldboard plow. However, a bare soil surface can be achieved with other tools, depending on the previous crop, amount of surface residue, and number and timing of tillage operations.

CURRENT ACREAGES OF CONSERVATION TILLAGE AND TRENDS

Several attempts have been made to extrapolate existing estimates of conservation tillage into the future. The first of these was prepared by the USDA Office of Planning and Evaluation (1975). This projection estimated that minimum tillage would be used on 95 percent of the U.S. planted cropland acres by 2010. The 1975 USDA projection has received considerable attention. Much of this attention has been a concern that it presents an overly optimistic outlook for the adoption of conservation tillage.

Pierre Crosson (1981), in his publication entitled, "Conservation tillage and conventional tillage: A comparative assessment," made the following statement, "...without pretending that there exists a solid base for it, I believe that economic factors could easily induce farmers to adopt conservation tillage on 50 to 60 percent of the nation's cropland by 2010." Crosson later indicated that his estimate of 50 to 60 percent was conservative.

The Office of Technology Assessment (OTA, 1982) used the same criteria as was used in the 1975 USDA projection except they used an upper, long-term limit of minimum tillage adoption of 75 percent of the cropland planted rather than 100 percent as was used in the 1975 USDA estimates. OTA projected that 75 percent of U.S. cropland may have some form of conservation tillage by 2010.

There are several reasons why these estimates vary so greatly, but the key ones are:

1. Differences in definition of conservation tillage.

2. Lack of a reliable data base from which projections can be made.

The 1975 USDA projection of 95 percent by 2010 is not overly optimistic if one considers the definition of the practices being projected. This USDA projection was for minimum tillage and not necessarily conservation tillage. Minimum tillage was considered in the 1975 report as a concept or set of criteria for farm production methods. Of key importance in the 1975 definition of minimum tillage are the following statements: "The term conservation tillage means tillage that is consistent with maintenance of a protective cover of crop residue on the soil surface. Often this is minimum tillage, but sometimes it is not. Moldboard plowing for corn production in the Midwest could be necessary for adequate net income, and, therefore, it could be minimum tillage, but not conservation tillage." Minimum tillage in 1975, and as was then projected, meant simply "reducing soil manipulations to the minimum that was biologically, technologically, and economically feasible to the crop producer under particular soil and climatic conditions."

It is clear that before one projection can be compared to another or one estimate considered too optimistic or pessimistic, the definitions and assumptions under which the estimates were made must be thoroughly examined.

The Soil Conservation Service has estimated conservation tillage acreage in the period 1963–1985 in cooperation with CTIC (Table 1). In addition these acreages are plotted against years in Figure 1.

Magleby et al. (1985) summarized findings from a survey of the use of conservation tillage in 1983. They found the following: particularly high rates of adoption have occurred in the Corn Belt and Northern Plains; most conservation tillage is used in conjunction with single-crop corn, soybeans and small grains; farmers on mid-to-large sized farms and who rent land from others have the highest adoption; convervation tillage is used on some farms with fairly level terrain for reasons other than soil conservation such as savings in production costs; many farmers with steep cropland are not using conservation tillage even though it could provide significant erosion control; cost and time savings are about as important as soil and water conservation as reasons for using conservation tillage; most farmers who adopted conservation tillage in 1983 did so without cost-sharing assistance (it should be pointed out that 1983 was the PIK year and so might not be a fair assessment in this regard).

Table 1. Conservation Tillage - SCS On-the-Land Estimate[1]

Year	Acres	Annual Increase, Percent
1963	3,769,831	—
1964	4,994,500	32
1965	6,619,449	33
1966	8,094,973	22
1967	10,716,222	33
1968	12,360,228	15
1969	15,808,044	29
1970	18,582,842	18
1971	21,804,073	17
1972	24,073,162	10
1973	29,483,423	22
1974	32,630,025	11
1975	35,829,153	10
1976	39,161,170	10
1977	47,456,720	21
1978[2]	51,703,377	9
1979[2]	55,000,000	6
1980[2]	65,000,000	18
1981[2]	72,150,000	11
1982[2]	82,000,000	14
1983	78,000,000	-5
1984	92,000,000	18
1985	100,000,000	9

1 1963-1977 - Estimates from SCS reporting system, "On-the-Land" 99 report made by district conservationists; 1983-1985 from the Conservation Tillage Information Center (rounded to the nearest million acres).

2 These estimates based on SCS records for prior years.

1985
NATIONAL SURVEY
CONSERVATION
TILLAGE PRACTICES

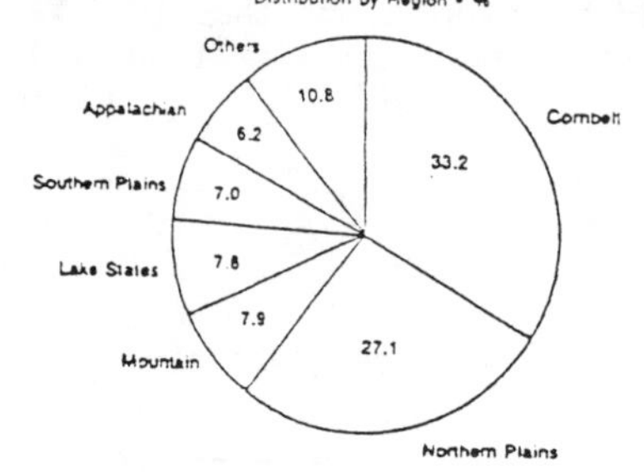

NO TILL

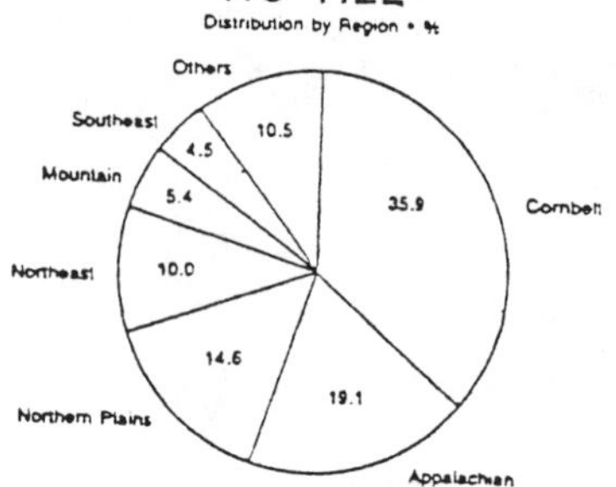

RIDGE TILL

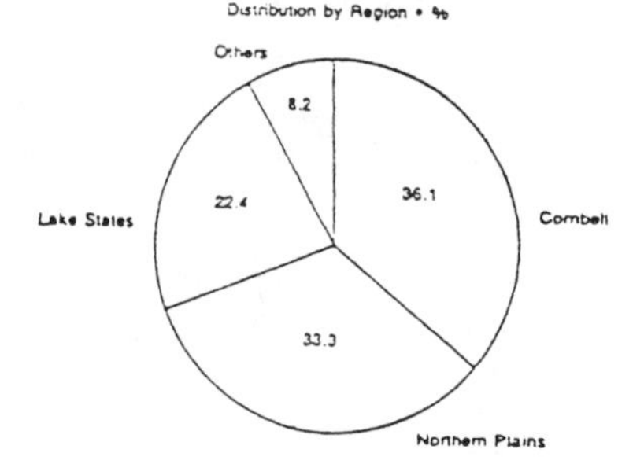

STRIP TILL

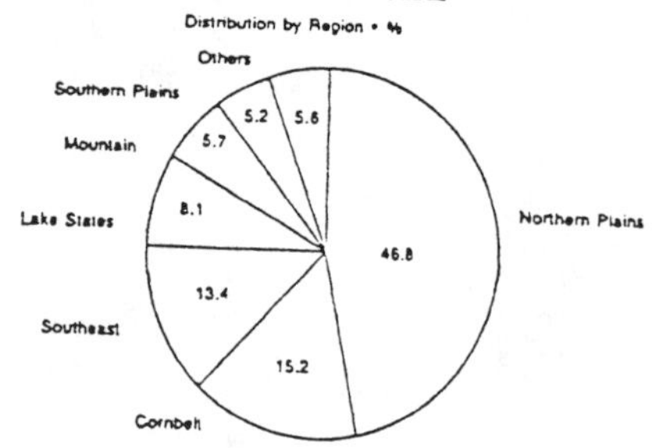

MULCH TILL

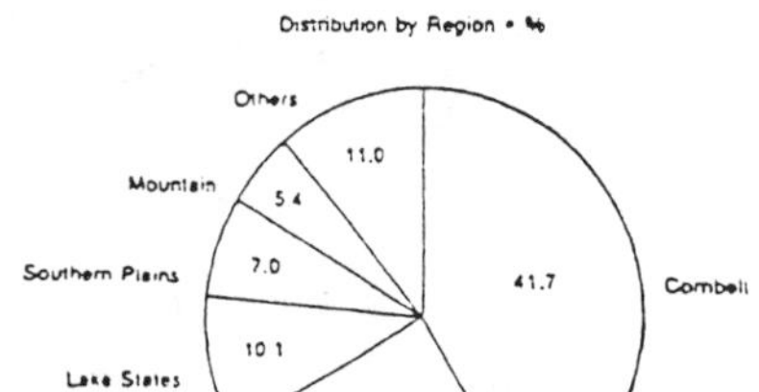

REDUCED TILL

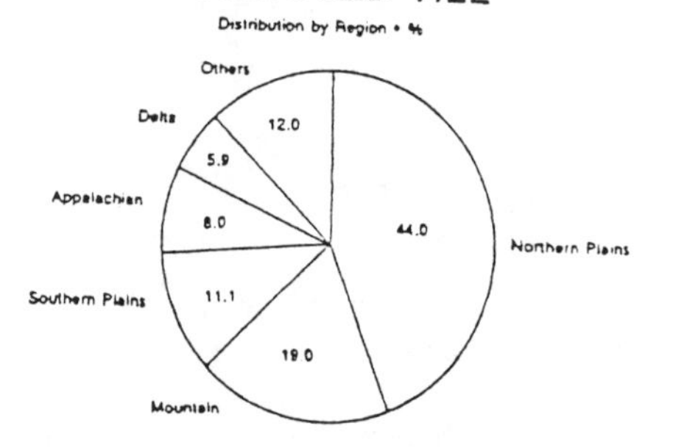

Figure 1. Estimates of U.S. cropland acres in conservation tillage 1963 through 1985. (See Table 1 for data source)

1985 Estimates

The data presented in Table 2 are extracted from the fourth National Survey of Conservation Tillage Practices, published by CTIC (1985). Conservation tillage adoption continues to expand with 31.43% of all crops planted in 1985 with some form of conservation tillage. The distribution of conservation tillage by regions on a percentage basis is shown in Figure 2.

Table 2. 1985 Acres by Various Conservation Tillage Types (CTIC, 1982-85).

		Tillage System					
Region	Acres	No-till	Ridge	Strip	Mulch	Reduced[1]	Cons.[2]
Appalachian	15,989,290	2,859,704	45,213	26,962	1,786,510	1,502,445	6,220,834
Caribbean	76,024	1,608	0	0	27,299	11,576	40,483
Corn Belt	80,771,806	5,371,066	708,467	95,427	26,409,731	533,601	33,118,301
Delta	17,032,181	281,876	2,466	105	551,698	1,099,783	1,935,928
Lake States	33,239,907	617,096	439,467	50,698	6,401,617	208,942	7,717,820
Mountain	24,172,206	803,131	40,184	35,252	3,421,705	3,562,916	7,863,188
Northeast	9,211,399	1,496,433	2,169	2,250	1,341,830	614,109	3,456,791
Northern Plains	71,144,064	2,186,603	633,061	291,754	15,669,686	8,257,673	27,058,777
Pacific	15,871,055	266,293	11,310	4,490	1,609,208	664,788	2,556,089
Southeast	14,839,685	677,942	10,385	84,115	1,674,819	197,447	2,644,708
Southern Plains	34,515,214	388,143	50,201	32,050	4,415,729	2,085,193	6,971,316
Nation	316,862,831	14,949,895	1,962,923	623,103	63,309,832	8,738,482	99,584,235

1 Total acres planted in all tillage types. Does not include permanent pasture, fallow, or conservation use.
2 Sum of No-till, Ridge-till, Strip-till, Mulch-till, Reduced-till

In comparing the 1984 data to that of 1985, a few observations are in order:

Nationally

- There were more acres planted in some form of conservation tillage than ever before - a total of 99.6 million acres or an increase of 2.8 million acres.

- With respect to tillage type:

  - No-till increased by 700,000 acres.
  - Ridge-till realized a 45% rate of growth.
  - Strip-till declined by 50,000 acres.
  - Mulch-till, accounting for 64% of all conservation tillage, grew by 6.7 million acres.

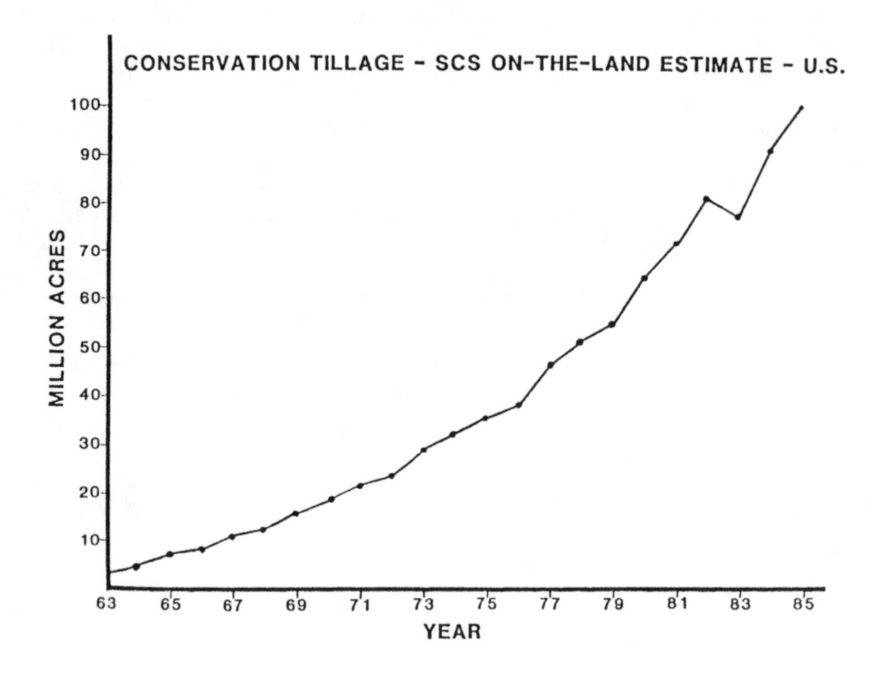

Figure 2. Distribution of conservation tillage systems by regions. Source: 1985 National Survey of Conservation Tillage Practices. (CTIC 1985)

- With respect to crops:

  - Full season corn and soybeans realized the largest increases in conservation tillage.
  - The total acres "planted" for both conventional and conservation tillage decreased by 10.4 million acres.

Regionally

- The Corn Belt and Northern Plains account for over 60% of all conservation tillage.
- Over 75% of the increase in no-till acreage came from the Corn Belt and Mountain regions.
- Over 75% of the increase in ridge-till acreage came from the Corn Belt and Northern Plains.

COMPARISON OF CONSERVATION TILLAGE AND CONVENTIONAL TILLAGE

The principal objectives of this workshop were to evaluate the environmental impacts of conservation tillage on surface and ground water quality. Since one of the major effects of "conservation tillage" compared to conventional tillage is to reduce soil and/or water loss, it might be well to discuss those tillage induced factors that have significant impacts. Major factors would include surface residue cover in relation to timing of erosive weather conditions, surface roughness, soil porosity, and pore-size distribution. The most significant factor affecting soil and water loss is crop residue cover. We will discuss the contrast between conservation and conventional tillage as to their effects on percent cover. Included will be the interactions between tillage system and crop species, and the effect of extent and timing of tillage on percent cover. The other effects of conservation tillage on hydrology are discussed in a later chapter.

Percent Cover

Tillage System and Crop Interaction

Percent residue cover over much of the U.S. where rowcrops are grown is most critical for erosion control from the seedbed establishement period until significant crop canopy is attained. Many researchers estimate residue cover shortly after planting to represent this period. (Remember this is the time frame identified in the definition of conservation tillage.) For drilled small grains, the most critical need for surface residue cover occurs after-harvest if surface cover is reduced by tillage until the new crop provides canopy protection. This period would vary appreciably depending on whether the crop was a fall or spring seeded cereal or whether it was summer fallowed.

In an attempt to illustrate the interactions of different crop types on percent surface cover, Griffith et al. (1986) developed the information shown in Table 3 which is a summary of several studies and is intended to represent "typical" situations. The data show major effects both for tillage systems and crop species. The tillage system - crop specie interaction is very pronounced with soybeans. Note that many of the tillage systems that we would normally think of as "conservation tillage" do not meet the 30% cover "after plant" criteria. Griffith points out, however, that the values shown for the no-plow systems could easily vary ± 30% due to management of the crop (crop-yield) and tillage system.

Table 3. Residue Cover after Planting for Typical Tillage vs. Previous Crop Combinations. (Griffith et al., 1986)

	Previous Crop			
Tillage system	Corn	Soy-beans	Small grain	Sod
	% cover			
Moldboard plow, disk, field cultivate	5	2	5	10
Chisel (10 cm twisted points)				
disk twice	15	2	10	20
disk 1, field cultivate 1	20	5	15	25
Chisel (5 cm straight points)				
+ disk twice	20	5	15	25
+ disk 1, field cultivate 1	30	10	25	30
Primary tillage disk (deeper than 15 cm)				
+ disk twice (standard tandem)	10	5	15	20
+ disk 1, field cultivate 1	20	10	20	—
Shallow disking (less than 15 cm)				
once (standard tandem)	40	20	45	50
twice (standard tandem)	20	10	25	30
Field cultivate once	—	30	—	—
Till-plant in ridge	30	20	—	—
In-row subsoil, plant	70	50	80	85
No-tillage plant	80	60	90	95

Seasonal Effect

The amount of crop residues on the soil surface is also affected by seasonal changes during the crop year as well as crop specie and the timing and type of tillage. These effects are illustrated by the work of Cruz (1982) in Table 4. Note particularly the major change in surface cover when soybean residue was chiseled compared to chiseled corn residue. These data show that the only tillage system of the three to fully meet the residue qualifications after plant would be no-till.

Table 4. Surface residue cover as affected by seasonal change, crop specie and timing and type of tillage. (Cruz, 1982)

Prior Crop[1]	Tillage	Seasonal Period: After Harvest	After Fall Plow	Before Secondary Spring Tillage	After Planting[2]
		% Cover			
Corn	Conventional	98	1	1	2
	Chisel	97	56	42	22
	No-till	99	99	85	62
Soybeans	Conventional	88	1	1	1
	Chisel	89	20	20	6
	No-till	95	95	72	33

1 After six years of corn-soybean rotation.

2 Secondary tillage for the conventional and chisel treatments consisted of two trips - one disking followed by a field cultivator. Anhydrous NH_3 was applied prior to the after plant measurement and explains why no-till values are further reduced.

Degree of Tillage

The degree of intensity of tillage can also have significant effects on the amount of surface cover. Allen and Fenster (1986), working with small grain residue, developed the information in Table 5.

These data essentially show that, the more intensive the tillage, the greater the reduction in surface residue. The V-sweep and rod weeder have only minor effects while systems such as the deeper disk operation greatly reduce surface cover. In most cases several of these operations will occur from harvest to small grain seeding which, when combined, greatly reduce surface cover.

Table 5. Average residue reduction per tillage operation (Allen and Fenster, 1986).

Tillage Operation	Residue Reduction per Tillage Operation
	%
V-sweeps, 76 cm or larger	10
Chisel plows	25
Rodweeder	5 - 10
Tandem or offset disk	
operated 7.5 cm deep	30
operated 15 cm deep	70

EROSION/SEDIMENT REDUCTION POTENTIAL

The sheet and rill erosion reduction benefit of conservation tillage varies between 50 to 90 percent compared to erosion under conventional tillage as illustrated by studies with simulated and natural rain (Tables 6 and 7, respectively). This dramatic reduction in erosion is primarily a result of the residue cover on, and incorporated into, the soil surface. As a general rule, 30 percent surface residue cover will provide 50 percent reduction in sheet and rill erosion. A 30 percent residue cover is often retained using existing chisel-type tillage equipment. However, to obtain a 90 percent reduction in sheet and rill erosion, no-till is generally needed along with high residue-producing crops or a cover crop.

Although this dramatic reduction in sheet and rill erosion is well documented in the literature, the corresponding effect of conservation tillage on sediment reduction is not. It may seem logical to assume that a 50 percent reduction in sheet and rill erosion would also have a similar reduction in sediment yield. However, most sediment leaves a field via concentrated flow. If conservation tillage reduces the amount of soil entering concentrated flow patterns, the detachment capacity in the concentrated flow itself may increase and erode an appropriate amount of soil from the channel. The end result may be less gross erosion on a field, but without a corresponding reduction in sediment delivery. However, conservation tillage will leave a significant amount of surface residue over the whole field. The residue not only intercepts the falling raindrop, but also slows runoff. As the rate of runoff is reduced, greater infiltration generally occurs and less runoff results. As less runoff occurs in concentrated flow channels, less sediment is carried from the field. Therefore, conservation tillage can have a very favorable impact not only on sheet and rill erosion, but also on concentrated flow erosion and resulting sediment delivery.

Table 6. Surface Cover and Soil Loss from Various Tillage Systems on 5% Slope Land Tilled on the Contour Following Corn and Soybeans (Siemens and Oschwald, 1986).

Tillage System	Surface Cover following Corn	Surface Cover following Soybeans	Soil Loss following Corn	Soil Loss following Soybeans
	%	%	--- MG/ha ---	
Fall Moldboard Plow	4	2	12.8	25.5
Fall Disk-Chisel	50	11	1.3	7.4
No-Till	85	59	1.1	3.8

1 Catlin silt loam with slope length 10.6 m. Tests made using 127 mm of simulated rainfall after over-winter weathering, put prior to any spring tillage.

Table 7. Runoff and Soil Loss on Contour-Row Fields in Corn Watersheds-Coshocton, Ohio[1]. (Harrold and Edwards, 1972).

Tillage System	Slope %	Runoff % of Rain	Soil Loss (Mg/ha)
Moldboard Plow	5.8	42	7.21
No-till	20.7	49	.07

1 About 135 mm of natural rain fell within a 7-hour period early in the growing season.

A modeling approach to determine the effect of conservation tillage on sediment yield was performed using 18 sample watersheds (1% sample of the total basin) in the western basin of Lake Erie (Beasley et al. 1985). This ANSWERS model predicted that if all cropland in the basin was no-tilled, sediment yield would be reduced 80%. However, a more likely scenario using chisel tillage and ridge tillage would result in sediment yield reductions of 15 to 45%.

Another major concern is that if conservation tillage increases infiltration, what effect might this have on deep percolation causing transport of pollutants to the groundwater? Research in this area is discussed in a later chapter.

REFERENCES

Allen, R. R. and C. R. Fenster. 1986. Stubble mulch equipment for soil and water conservation in the Great Plains. J. Soil and Water Cons. 41:11-16.

Beasley, D. B., E. J. Monke, E. R. Miller, and L. F. Huggins. 1985. Using simulation to assess the impacts of conservation tillage on movement of sediment and phosphorus into Lake Erie. J. Soil and Water Cons. 40:233-241.

Conservation Tillage Information Center. 1982-85. National Surveys Conservation Tillage Practices. Published by CTIC, Ft. Wayne, IN.

Crosson, Pierre. 1981. Conservation tillage and conventional tillage: A comparative assessment. Soil Cons. Soc. of Amer., Ankeny, IA.

Cruz, J. C. 1982. Effect of crop rotation and tillage system on some soil physical properties, root distribution and crop production. Ph.D. Thesis, Purdue Univ., W. Lafayette, IN.

Griffith, D. R., J. V. Mannering, and J. E. Box. 1986. Soil and moisture management with reduced tillage. In No-tillage and surface tillage agriculture. M. A. Sprague and G. B. Triplett (ed.) Wiley and Sons, NY. pp. 19-57.

Harrold, L. L. and W. M. Edwards. 1972. A severe rainstorm test of no-till corn. J. Soil and Water Cons. 27.

Magleby, R. D. Gatsby, D. Colacicco, and J. Thigpen. 1985. Trends in conservation tillage use. J. Soil and Water Cons. 40:274-276.

Mannering, J. V. and C. R. Fenster. 1983. What is conservation tillage? J. Soil and Water Cons. 38:140-143.

Office of Technology Assessment. 1982. Impacts of technology on U.S. cropland and rangeland productivity. Congressional Board of the 97th Congress. Library of Congress Catalog Card No. 82-600596. Washington D.C. 266 p.

Resource Conservation Glossary. 1982. Third Edition. Soil Cons. Soc. of Amer., Ankeny, IA.

Siemens, J. C., E. C. Dickey, and E. D. Threadgill. 1985. Definitions of tillage systems for corn. National Corn Handbook, CES, Purdue Univ., West Lafayette, IN 4pp.

Siemens, J. C., and W. R. Oschwald. 1976. Corn-soybean tillage systems: Erosion control, effects on crop production, costs. Amer. Soc. Agr. Eng. Paper No. 76-2552. St. Joseph, MI.

USDA Office of Planning and Evaluation. 1975. Minimum Tillage: A Preliminary technology Assessment.

CHAPTER 2

OVERVIEW OF PEST MANAGEMENT FOR CONSERVATION TILLAGE SYSTEMS

R. S. Fawcett,
Iowa State University, Ames Iowa

INTRODUCTION

Changes in tillage practices can significantly affect pest problems, thus changing pest management strategies and pesticide use. Concerns have been raised by some observers that a trade off may occur with the adoption of conservation tillage. The beneficial result of reduced soil erosion with conservation tillage may come at the expense of increased pesticide use which may increase water contamination potential (Hinkle, 1983). This paper will address the implications of conservation tillage on weeds, insects, and plant pathogens and on projected changes in pesticide use. Greatest emphasis will be placed on weeds, as extensive research has been conducted on the impact of tillage on weeds and herbicides, and herbicide use accounts for about 85% of total pesticide use in the United States (Delvo, 1984). Corn and soybeans account for 80-85% of present herbicide use.

Terminology referring to types of tillage systems is not uniform in the literature. In this paper, terminology established by the Conservation Tillage Information Center will be used. Conventional tillage refers to systems which totally disturb the soil surface and bury residue from the previous crop. The moldboard plow has traditionally been used as the primary tillage tool in these systems. Mulch tillage refers to systems which disturb the total soil surface but leave a minimum of 30% of the soil surface covered with crop residue. Tillage tools used in this system include the chisel plow, disk, and

Effects of Conservation Tillage on Groundwater Quality: Nitrates and Pesticides, Terry J. Logan et al., eds. © 1987 Lewis Publishers, Inc., Chelsea, Michigan 48118. Printed in USA.

field cultivator, among others. Mulch tillage systems are often referred to as reduced tillage in the literature. No-till systems leave the soil undisturbed prior to planting. Planting is completed in a narrow seedbed approximately 3 to 8 cm wide. Ridge-till systems leave the soil undisturbed prior to planting. Approximately one third of the soil surface is tilled at planting with sweeps or row cleaners. Planting is completed on ridges usually 10 to 15 cm higher than row middles. Cultivation is used to rebuild ridges.

WEEDS

Repeated tillage has historically been the primary method of weed control in row crops. Tillage kills existing vegetation and creates an even start situation for the crop and weeds which must begin growth from seeds or vegetative propagules. Most weeds commonly associated with row crops are spring annual species favored by disturbed soil conditions. Weed seeds respond to physical or chemical changes accompanying tillage and are stimulated to germinate. Seeds may respond to light (Taylorson, 1972; Taylorson and Hendricks, 1972; Wesson and Wareing, 1969), to increases in oxygen concentration (Holm, 1972), to temperature increases (Taylorson, 1972; Taylorson and Hendricks, 1972), or other factors such as drying or mechanical scarification (La Croix and Staniforth, 1964). In the past, tillage operations were often repeated in an effort to deplete the number of viable weed seeds in the soil. Weed seedlings germinating in response to tillage were killed by a second tillage operation. This procedure was repeated several times prior to crop planting.

Tillage also changes the distribution of weed seeds in the soil (Pareja et al., 1985; Wicks and Sommerhalder, 1971). Seeds of most weed species germinate from the top 1 or 2 cm of soil (De La Cruz, 1974). Weed seeds placed several cm deep in the soil by tillage often become more dormant and do not germinate until brought near the soil surface by later tillage (Taylorson, 1970). Thus plowing down weed seeds can delay, but not necessarily eliminate, later weed germination. Weed seeds placed near the soil surface have a shorter longevity than deeper weed seeds, because they either germinate or decay more quickly than deeper weed seeds (Taylorson, 1970). Pareja et al. (1985) have shown that tillage places more weed seeds inside of soil aggregates. This has the consequence of increasing weed seed dormancy and reducing germination, probably due to lower oxygen availability within the aggregates (Pareja and Staniforth, 1985). In contrast, as tillage is reduced, more weed seeds are located either outside soil aggregates or in small soil aggregates where they more readily germinate. Shallow placement of weed seeds either outside of soil aggregates or in small aggregates favors weed seed germination.

Thus reduction in tillage can have the effect of increasing the germination of newly produced weed seeds over the short term. On the other hand, the reservoir of dormant weed seeds located below the top few cm of soil is not transferred to near the soil surface if deep tillage is not performed. For these reasons, annual weed populations can increase where conservation tillage is performed in fields where a large number of weeds were uncontrolled and produced seed the previous growing season (Lugo, 1984). However, weed populations can decline due to depletion of shallow weed seeds in fields where conservation tillage follows effective weed control practices and weed seed production is minimal (Fawcett, 1985; Schaefer, 1984). Adequate weed control is thus very important during the first few years after fields are converted to conservation tillage, as success or failure can have a great impact on future weed populations.

Changes in tillage practices may favor certain weed species over others. Many annual grass species such as giant foxtail (Setaria faberi Herrm.), fall panicum (Panicum dichotomiflorum Michx.), and crabgrass [Digitaria sanguinalis (L.) Scop.] can germinate from very shallow soil depths and are well adapted to conservation tillage systems (Becker, 1978; Fawcett, 1985). One study showed that giant foxtail populations were similar in plowed and no-till plots, but were increased 30% by shallow tillage (Becker, 1978).

Populations of certain large-seeded broadleaf weed species such as velvetleaf (Abutilon theophrasti L.) and cocklebur (Xanium strumarium L.) are often decreased by reductions in tillage (Becker, 1978; Schaefer, 1984). Species which germinate under cool soil conditions such as common lambsquarters (Chenopodium album L.) and Pennsylvania smartweed (Polygonum pennsylvanicum L.) may be prevalent in no-till fields prior to crop planting, as their early germination allows them to out-compete later germinating weeds. On the other hand, with mulch tillage or conventional tillage, the early germinating species may be eliminated by preplanting tillage, favoring weeds germinating under warmer soil conditions, such as redroot pigweed (Amaranthus retroflexus L.).

Winter annual weeds such as horseweed [Conyza canadensis (L.) Cronq.] are killed by preplanting tillage in reduced and conventional tillage systems for row crops. Such winter annual weeds can persist in no-till systems and require herbicide treatment. In the Great Plains the winter annual, downy brome (Bromus tectorum L.) became a significant weed problem with the stubble mulch system in a winter wheat-fallow rotation (Wicks, 1985).

Vegetatively reproducing perennial weeds are found in all tillage systems. Repeated tillage may be effective in suppressing some perennial weeds by depletion of stored energy reserves thus inhibiting new shoot development. Reducing or eliminating tillage may allow certain perennial weeds to increase more rapidly. In a long-term Iowa tillage study,

Becker (1982) showed that hemp dogbane (*Apocynum cannabinum* L.) populations increased with all tillage systems in a corn-soybean rotation, but incresed most rapidly with no-till. Common milkweed (*Asclepias syriaca* L.) also increased more rapidly with no-till and ridge-till. Other broadleaf perennial weeds noted to have sometimes increased with reductions in tillage include Canada thistle [*Cirsium arvensis* (L.) Scop.], horsenettle (*Solanum carolinense* L.), and ground cherry (*Physalis* sp.) (Triplett and Lytle, 1972; Williams and Wicks, 1978). Vegetatively reproducing grass species such as quackgrass (*Agropyron repens* L.) and wirestem muhly [*Muhlenbergia frondosa* (Pair.) Fern.] also have tended to increase as tillage is reduced (Fawcett, 1985). Intensive tillage practiced in rotation with reduced tillage has not been successful in reducing existing stands of hemp dogbane. Three years of moldboard plowing did not reduce hemp dogbane populations in Iowa (Becker, 1982).

Simple perennial weeds and crops are killed by tillage and thus are usually not a problem in conventional and mulch tillage systems. They can persist, however, in no-till systems. Dandelions (*Taraxacum officionale* L.) and forage legumes such as alfalfa (*Medicago sativa* L.) and clover (*Trifolium* sp.) and grasses such as bromegrass (*Bromus inermis* Leyss.), orchardgrass (*Dactylis glomerata* L.) and tall fescue (*Festuca arundinacea* Schreb.) are examples of species which are usually controlled by herbicides in no-till systems. Woody species such as sassafras (*Sassafras albidum* L.), brambles (*Rubrus* sp.), and other small shrubs often also persist in no-till systems (Williams and Wicks, 1978). These species are killed by intensive tillage rotated with no-till.

The tillage system used may dictate herbicide application method options. Herbicides used in mulch tillage systems are often identical to herbicides used with conventional tillage. All application methods are feasible -- preplant incorporated, preemergence, and postemergence. Incorporation of herbicides in mulch tillage systems may require more skill than incorporation in conventional tillage systems due to roughness of surface soil and surface crop residue (Fawcett, 1985).

Incorporation of herbicides is not possible with no-till or ridge-till systems, so herbicide options are limited to preemergence and postemergence. If weeds have emerged prior to planting, a nonselective herbicide such as glyphosate or paraquat may be needed. Residual herbicides are often applied simultaneously. The use of nonselective herbicides in no-till systems varies greatly by region. In northern states often few weeds have emerged by corn planting, thus eliminating the need for the nonselective treatment when corn is planted into row crop stubble. Some residual herbicides have postemergence activity and can subsititute for nonselective herbicides in controlling small emerged weeds at planting. Cyanazine and atrazine are often used to control small emerged weeds when planting no-till and provide residual control as well. Liquid fertilizer carriers are also useful to control small emerged vegetation in the absence of nonselective herbicides. Linuron and metribuzin control small emerged weeds as well as provide residual control when planting soybeans. The early preplant system of herbicide application also usually eliminates the need for nonselective herbicides when no-till planting into row crop residue. Residual herbicides are applied several weeks prior to corn or soybean planting and weed germination. Besides eliminating the need for nonselective herbicides, early application often improves herbicide performance due to more favorable rainfall patterns for herbicide activation (Fawcett et al., 1983).

In southern states or with later plantings, weeds are taller at planting time, and there is more likelihood that nonselective herbicides will be needed in no-till systems. Early preplant application systems are also less adaptable to southern regions. No-till planting into sod often requires a nonselective herbicide due to vigorous growth of forage grasses. When double crop planting soybeans following small grains, nonselective treatments are usually required to control existing weed growth. Nonselective herbicides are usually needed when planting into cover crops.

Ecofallow or chemical fallow systems practiced in the Great Plains utilize postemergence and/or residual herbicides to control weeds during the fallow season instead of repeated tillage. This system reduces erosion and water evaporation loss. These herbicides would not be necessary in conventional fallow systems using tillage for weed control.

Crop residue left on the soil surface by conservation tillage can intercept soil-applied herbicides and reduce the amount of herbicide reaching the soil surface. Most herbicides are readily washed from crop residue to the soil by rainfall, but there are differences between herbicides and types of residue. Martin et al. (1978) found that when ^{14}C-herbicides were applied to corn stalk residue, most of the applied atrazine, cyanazine, alachlor, and propachlor washed off with simulated rainfall. Approximately the same amount of herbicide washed off with the first 0.5 cm of water as did with the next 3.0 cm of water. Ghadiri et al. (1984) determined that 60% of applied atrazine was intercepted by standing wheat stubble in the field. After 3 weeks and 50mm of rainfall, 90% of the atrazine washed from the wheat stubble, increasing soil concentrations twofold. Thus the effect of crop residue on herbicides is more to delay their reaching the soil surface rather than to prevent it. The ultimate distribution of herbicide can be changed by heavy residue. Postemergence herbicides are not influenced by crop residue.

Some researchers have reported that crop residue reduced the activity of residual herbicides. Banks and Robinson (1983) found that increasing wheat straw levels reduced the efficacy of alachlor, acetochlor, and metolachlor. Others have found that, while increasing rates of crop residue can reduce the amount of herbicide initially reaching the soil surface, efficacy has been satisfactory when labelled rates of herbicides are used (Fawcett, 1985). Erbach and Lovely (1975) found that corn stalk residue (6200 kg/ha) did not reduce efficacy when recommended rates of alachlor and atrazine were used, but reduced rates of application were adversely affected by corn stalk residue.

Crop residue mulches may reduce weed seed germination, thus offsetting possible reductions in herbicide efficacy due to retention of herbicide on crop residue. Banks and Robinson (1983) found that wheat straw mulch had more effect on controlling spiny amaranth (Amaranthus spinosus L.) and tall morningglory [Ipomoea purpurea (L.) Roth] than preemergence herbicides used in nonmulched areas. Thilsted and Murray (1980) also found better control of pigweed (Amaranthus sp.) in untreated plots with straw cover than in untreated bare soil plots. Crutchfield et al. (1986) found that, although wheat straw intercepted part of the applied metolachlor, there were still fewer weeds in mulched plots than unmulched plots. They concluded that increasing the herbicide rate was not necessary to maintain adequate weed control in no-till winter wheat stubble since the mulch itself provided some measure of weed control. Some of the weed control properties of crop residue mulches may be due to allelopathic compounds. Steinsiek et al. (1982) found that aqueous extracts of wheat straw inhibited the germination of four of the six weed species studied.

Residual herbicide activity may be changed over the long-term by soil property changes caused by tillage system. Soil organic matter may increase with time or decrease more slowly under no-tillage compared to moldboard plowing, especially near the soil surface. Fleige and Baeumer (1974) reported that, after 5 years organic carbon content of the 0- to 5-cm layer of soil was 50% greater for no-tillage than for plowing. Slack et al. (1978) reported a similar increase in organic matter content for the top 8 cm of soil after 6 years of no-tillage production. Higher organic matter could reduce herbicide activity due to greater herbicide adsorption.

Efficacy and persistence of triazine herbicides is influenced by soil pH. In no-till systems, surface soils can become acidic due to surface applications of nitrogen fertilizers (Griffith et al., 1977; Triplett and Van Doren, 1969). This acidity can reduce the persistence and efficacy of atrazine and simazine (Kells et al., 1980; Schmaffinger et al., 1977; Slack et al., 1978). Efficacy can be improved in such situations by the addition of agricultural lime.

The label directions of most residual herbicides do not describe using higher rates where surface crop residue is present. Efficacy data generated to support product registrations have shown that performance of most herbicides is satisfactory at labelled rates in all tillage systems. One herbicide label that specifically addresses rate variations with surface crop residue is the cyanazine label. It states "Where heavy crop residue exists the Bladex rate should be increased by 25%." Cyanazine has been shown to be retained more strongly by crop residue than some other herbicides (Martin et al., 1978). A 1984 survey of the Lake Erie Drainage Basin showed that conservation tillage farmers did not use higher rates of individual herbicides than conventional tillage farmers (Christensen et al., 1985). For example, average application rates of atrazine in corn were 3.0, 2.6, 2.7, 2.6, and 2.9 kg/ha for no-till, ridge-till, chisel plow, disk, and conventional tillage systems, respectively. Use rates of alachlor were 2.4, 2.2, 2.4, 2.5 and 2.5 kg/ha for no-till, ridge-till chisel plow, disk, and conventional tillage systems, respectively.

Certain herbicide labels allow a range of rates for specific soil types while others describe only one rate for each soil type. Although most herbicide labels do not describe using higher rates in conservation tillage than in conventional tillage, some Extension specialists may recommend using herbicide rates at the top of label rate ranges with no-till or in other situations where surface crop residue is heavy. Because successful weed control and prevention of weed seed production in the first years after conversion to no-till can have a large impact on future weed control, this recommendation may have merit. Fields with histories of poor weed control and thus high quantities of weed seed in surface soil may require higher herbicide rates for satisfactory weed control. After several years of satsifactory weed control in no-till systems, there is less merit for the use of elevated herbicide rates.

Efficient application of herbicides can reduce the necessity for elevated herbicide rates. Application systems which delivery more spray droplets per unit area have improved the performance of residual herbicides under heavy surface crop residue conditions (Fawcett and Owen, 1984). The nonselective herbicide paraquat requires relatively high spray volumes and complete coverage of weed foliage for best results, while glyphosate has been found to be most active when applied in water volumes less than 93 liters/ha (Fawcett, 1985). Thus better education of pesticide applicators to improve application techniques can improve herbicide efficacy and in some cases reduce necessary rates of application.

The greatest impact of conservation tillage on herbicide use may be in changes in the kinds of herbicides used as dictated by changes in weed problems. The use of herbicides which must be incorporated could be reduced if no-till and ridge-till planting increases. It is unlikely that total kg of herbicide used per hectare will increase greatly. This is especially true considering the introduction of new herbicides active at very low rates. Herbicides used with mulch tillage systems are often identical to herbicides used with conventional tillage systems. There is little reason for increases in herbicide use to take place with these systems. It is with no-till (comprising about 6% of corn and 7% of soybeans in the United States) that changes in herbicide use patterns and possible increases in rates are most likely. It is important to note that nearly all corn and soybeans are currently treated with herbicides no matter what tillage system is used. Duffy (1983) reported that in 1982, 98, 97, and 93% of corn acreage was treated with herbicides for no-till, reduced-till (mulch-till), and conventional-till, respectively. For soybeans, 97, 95, and 92% of acreage was treated for no-till, reduced-till, and conventional-till, respectively. The adoption of herbicides by U.S. farmers preceded the adoption of conservation tillage. Figure 1 shows the adoption of the use of herbicides in Iowa corn production compared to the adoption of conservation tillage (as measured by nonuse of the moldboard plow). In 1968, when nearly all corn hectares were still moldboard plowed, already 75% of corn hectares were treated with herbicides. Over 90% of corn hectares were treated with herbicides in 1976 when only about 35% of hectares were not plowed. Thus the continued adoption of conservation tillage should not cause a greater percentage of corn and soybeans to be treated with herbicides than would have been treated in the absence of conservation tillage.

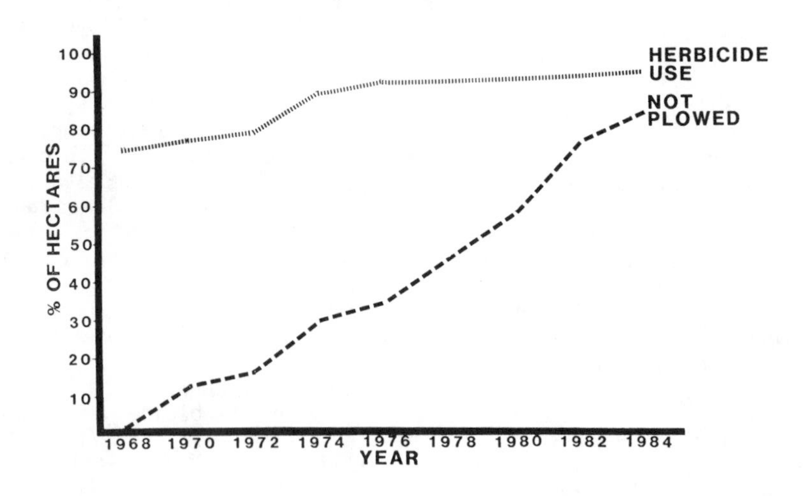

Figure 1. The adoption of the use of conservation tillage as measured by nonuse of the moldboard plow and the use of herbicides for corn production in Iowa. Herbicide use data are from Wallaces Farmer chemical use surveys and tillage data are from USDA SCS.

Some growers have reduced the amount of herbicide applied in the ridge-till system compared to conventional tillage. Because inter-row cultivation is an integral part of the ridge-till system, many growers now apply herbicide only in a band over the row, thereby reducing total rates by one half or more. Because tillage is performed at planting in the crop row with this system, nonselective herbicides are also often unnecessary.

Inter-row cultivation is still practiced by many no-till and mulch tillage growers. Duffy (1983) reports that in 1982 mechanical cultivations for weed control in the United States averaged 0.4, 1.4, and 1.5 per season for corn growers for no-till, reduced-till (mulch till) and conventional-till, respectively. Results for soybeans were 0.6, 1.7 and 2.3 for no-till, reduced-till, and conventional-till, respectively.

Surveys of pesticide use by tillage practice have shown either no significant increases or small increases in herbicide use with conservation tillage. Hanthorn and Duffy (1983) surveyed pesticide use by U.S. corn farmers using various tillage systems in 1980 and found that there were no statistically significant differences in rates of herbicide (the sum of all herbicide treatments) used in no-till, reduced-till (mulch till), and conventional-till. Cost of herbicide was greater for no-till, however ($42.58/ha for no-till compared to $31.37 for reduced-till and $28.13 for conventional-till). This illustrates the fact that, while conservation tillage may result in the use of different (and sometimes more costly) herbicides, it does not necessarily dictate the use of more herbicide. In the same survey, there was no significant increase in herbicide use in soybeans with conservation tillage for the Southeast and Midsouth regions, but no-till soybean growers in the Midwest did use significantly higher herbicide rates.

A 1984 survey of the Lake Erie Drainage Basin showed that, on average, corn growers practicing no-till and ridge-till used 18 and 1% more herbicide, respectively, than with conventional tillage. Soybean growers used 23% more herbicide in no-till and 17% more herbicide in ridge-till than in conventional tillage (Christensen et al., 1985). Many growers in this project were first-time users of conservation tillage. This may have influenced herbicide use.

Herbicide use may be higher for the first few years after conversion to no-till or ridge-till due to farmers' fears of weed problems or due to a learning process. A five-year study conducted in Iowa monitored all inputs into corn and soybean production under no-till, mulch-till, and conventional-tillage (Colvin et al., 1985). The resident farmer made all management decisions. Due to numerous unanticipated problems, the cost of soybean herbicides in the first year of no-till was $201.97/ha. In the last year of the study the soybean herbicide cost per hectare was $68.59, $66.52, and $66.52 for no-till, mulch-till, and conventional-till, respectively. Costs for corn herbicides per hectare were $48.91, $44.34 and $44.34 for no-till, mulch-till and conventional-till, respectively.

There may be greater possibility for average increases in rates of herbicide applied with the adoption of conservation tillage in small grain production. This is because some acreage not ordinarily treated in conventional tillage may receive herbicide with conservation tillage. Duffy (1983) reports that in 1982, percent of herbicide-treated acreage of wheat in the United States was 54, 28, and 47% for no-till, reduced-till (mulch-till), and conventional-till, respectively. For oats, herbicide treated acreage was 11, 14, and 32% for no-till, reducted-till, and conventional-till, respectively. Thus there presently is no clear pattern of greater treatment with conservation tillage.

There is no doubt that the adoption of conservation tillage will change weed problems and change herbicide use patterns. There may be increases in rates of herbicide with some systems. However, overall increases in herbicide use by present conservation tillage farmers are small or nonexistent. The properties of specific herbicides used and any direct effects that conservation tillage has on the ability of herbicides to run off or leach from treated sites appear to be more important water quality concerns than the quantity of herbicide applied.

INSECTS

Certain insect pests have been noted to be more prevalent with conservation tillage, while others are either not directly affected by tillage or have decreased with conservation tillage. Most cases of increases of insect pests have been in association with no-till. Herbicides which control emerged vegetation at planting can eliminate the natural hosts of some insects, forcing them to feed on the cultivated row crop (Gregory and Musick, 1976).

Small grains and cover crops, especially rye, attract armyworm (Pseudaletia unipuncta Haworth) moths for oviposition (Musick and Petty, 1974). Armyworm problems have been frequent when corn is planted no-till into such cover crops. Insecticide treatments may be needed for armyworm control when infestations exceed economic thresholds.

Black cutworm (Agrotis ipsilon Hufnagel) moths are attracted to early spring weed growth (Busching and Turpin, 1976). Thus greater cutworm problems may be encountered in no-till fields where weed growth is left until planting, compared to mulch or conventional tillage systems where spring preplanting tillage eliminates weed growth. Musick and Petty (1973) observed that the black curworm in Ohio attacked approximately 15% of the corn plants in no-tillage fields, whereas in adjacent, conventionally tilled fields, only one percent were attacked. Early preplant herbicide application with no-till planting has sometimes reduced black cutworm and armyworm problems by preventing early season weed growth and thus reducing the attractiveness of fields to moths (Foster, 1986). Black cutworm can sometimes be controlled by soil-applied preventative insecticide treatments which are already in use for corn rootworm control. However, heavy infestations may require rescue treatments applied solely for cutworm control (Foster and Stockdale, 1984). The development of synthetic black cutworm pheromone has made it possible to conveniently and economically monitor for moth flights in the spring. Many states now have networks of pheromone traps, which when used along with degree day accumulation data can predict when significant cutworm feeding will occur. Scouting fields at the predicted times can then determine which fields should be treated with insecticides and which fields do not require treatment.

The stalk borer (*Papaipeme nebris* Guenee) has increased as an insect problem in no-till corn production (Foster, 1986). This insect overwinters in grasses. With conventional and mulch tillage systems it is usually confined to areas near fencerows and waterways with perennial grass vegetation. With no-till systems, stalk borer can be found throughout fields in areas where weedy grasses were not controlled the previous season. Stalk borer problems can be controlled by controlling grassy weeds the previous season. Also, early preplant herbicide applications in some years have reduced stalk borer problems, as preventing weed emergence causes some larvae to starve prior to emergence of corn (Foster, 1986). Insecticides are sometimes required for stalk borer control.

The majority of insecticides used on corn are for corn rootworm control when corn is planted following corn. Data on the effect of conservation tillage on corn rootworms are conflicting. Musick and Petty (1974) observed that no-till caused a 4-fold increase in the number of rootworm eggs compared to conventional tillage, but Musick and Collins (1971) found that survival of northern corn rootworm (*Diabrotica longicornus* Say) eggs and larvae was reduced under no-till. It took 4 times more eggs in no-till than in conventional tillage to give the same root injury rating. They concluded that the need for control of corn rootworms in no-tillage corn will remain essentially the same as for conventional-tillage corn. In contrast, Gray (1986) reported that northern corn rootworm and western corn rootworm (*Diabrotica virgifera* LeConte) did not show egg laying preferences for any particular system. However, egg survival of both species was greater in no-till systems than with mulch-till or conventional tillage.

Tillage system could influence the performance of soil-applied insecticides used for corn rootworm control. Insecticides requiring shallow incorporation for best activity may not perform as well in no-till systems where heavy surface residue interferes with adequate incorporation of granules. In-furrow application of nonphytotoxic insecticides may be useful in these systems. However, recent research in Iowa has shown that surface banding of corn rootworm insecticides is the most effective application method in no-till as well as conventional tillage (Foster, 1986). Rotation of corn with other crops continues to be a popular and effective nonchemical control for corn rootworms, as this insect is an obligatory pest of corn.

European corn borer (Ostrinia nubilatis Hubner) is a major pest of conventional tillage corn. Tillage systems are not expected to appreciably affect damage by this pest (Musick and Petty, 1973). The failure to destroy corn stalk residues, the overwintering site for the larvae, could influence this pest, but the magnitude of this effect is not known (Musick and Beasley, 1978). Control is currently achieved through genetic resistance and insecticide treatments.

Insects such as wireworms (many species in Elarteridae family) and white grubs (many species in Scarabaeida family) are more common pests when corn follows old established meadows. While these pests are present with all tillage systems, lack of mechanical disturbance with no-till and failure to expose these insects to parasites and predators may enhance their populations (Phillips, 1984). Certain insect problems have been reduced by conservation tillage. Wicks and Klein (1985) report that sorghum greenbugs (Schizaphis graminum Rondani) were attracted more to tilled fields than to fields planted in untilled wheat stubble.

It is likely that insecticide use may increase in certain no-till fields due to specific insect problems associated with weed or cover crop vegetation. Insecticide use in mulch tillage systems which eliminate vegetation with preplanting tillage should not differ greatly from insecticide use in conventional tillage systems. Better weed control or changes in the timing of weed control such as early preplant herbicide application can reduce the necessity for insecticide treatments in some tillage systems. The use of insect scouting and economic thresholds can eliminate unnecessary insecticide application.

Hanthorn and Duffy (1983) reported that, in a survey of 10 major corn producing states in 1980, insecticide use with reduced tillage (mulch tillage) was identical to use with conventional tillage, but use with no-till was significantly greater (1.89kg/ha compared to 1.22 kg/ha). Use of insecticides on soybeans was not different for any tillage system. In a 1984 survey of the Lake Erie Drainage Basin, insecticide use was no different for no-till, ridge-till and conventional-till systems in corn and soybean production (Christensen et al., 1985).

PLANT PATHOGENS

Potentially, conservation tillage could impact many diseases, as inoculum is often carried on plant debris. Burying crop residue with tillage has been a suggested control technique for many foliar diseases. Leaving plant debris on the surface or partially buried in the soil may allow numerous pathogens to overwinter, or survive until the next crop is planted, but conditions for biological control of plant pathogens may also be increased (Phillips, 1984).

Several foliar pathogens of corn are known to survive in corn residue in mulch or no-till systems. They include Helminthosporium turcicum and H. maydis, which induce northern and southern corn leaf blight, respectively (Boosalis et al., 1981; Boosalis et al., 1967); Phyllosticta maydis, which induces brown spot (Burns and Shurtleff, 1973); Collectotrichum graminicola which induces anthracnose (Boosalis et al., 1981; Phillips et al., 1980); and Cercospora zeae-maydis which induces gray leaf spot (Hilty et al., 1979; Roane et al., 1974). Most of these diseases are more severe with conservation tillage than when corn debris is buried by plowing, but anthracnose stalk rot is not controlled by plowing (White et al., 1979). However, Elliot (1976) found seedling corn diseases (northern corn leaf blight, yellow leaf blight and Helminthosporium leaf spot) to be equal under no-till, minimum-till, and conventional tillage.

Several pathogenic bacteria overwinter in plant debris on the soil surface, but not in buried plant debris. Psedomonas glycinia which induces bacterial blight in soybean (Daft and Leben, 1973) and Corynebacterium nebraskense which induces Gosses bacterial wilt in corn (Schuster, 1975) are examples.

Conservation tillage can reduce soil temperatures in spring and early summer and may slow seed germination and seedling development thus potentially enhancing damping off and root diseases. The use of seed treatments can be effective in combating this problem. In Arkansas, soybean seedling diseases induced by Sclerotium rolfsii and Rhizoctonia solani increased in double cropped plots compared with conventionally planted plots, but seed treatments controlled the disease (Cox et al., 1976). In Delaware, no-till increased populations of Fusarium species in the soil, but there were no differences in the incidence of wilt induced by F. oxysporum in soybean between no-till and conventional tillage (Ferrant and Carroll, 1979).

Wicks and Klein (1985) observed that wheat diseases have not increased in Nebraska despite 10 years of ecofallow farming where crop residues are left on the soil surface. They postulated that the dry weather conditions of this area reduced the expression of some diseases such as tan spot. They expressed concern that Cephalosporium stripe potentially could become a serious disease problem unless crop rotation was practiced or tolerant varieties developed.

Minimum tillage techniques have been shown to decrease the incidence of take all disease in wheat in some regions (Brooks and Dawson, 1968) while increasing it in others (Cook et al., 1978). Eyespot of wheat caused by Pseudoscercosporella herpotrichoides has also been reduced by minimum tillage (Brooks and Dawson, 1968; Cook and Waldher, 1977). Tillage system had no effect on sharp eyespot (Rhizoctonia solani) and brown foot rot caused by Fusarium sp. (Hood, 1965).

Incidence of stalk rot of grain sorghum induced by Fusarium monifiliforme was decreased dramatically by ecofallow (Doupnik and Bousalis, 1980). Conservation tillage systems have also been noted to reduce stalk rot in corn. Increased moisture availability with ecofallow and conservation tillage is the probable reason for this effect.

Little research has been done on the effect of tillage on nematodes. In Iowa, populations of plant parasitic nematodes in corn were usually greater in no-till ridge plots than in plowed plots (Thomas, 1978). Effects of tillage on soybean nematodes have not been reported.

The primary methods of disease control for most agronomic crops in the United States continue to be genetic resistance and crop rotation. Use of foliar fungicides is not common, although as new effective and economical treatments are developed, their use could increase. Duffy (1983) reported that in 1982 less than 1% of corn, 1% of soybeans, and 2% of wheat received a foliar fungicide treatment. Fungicide use was not increased with conservation tillage. Fungicide seed treatments are used universally for corn. Such seed treatments are not as common with soybeans. Because of increased likelihood of seedling damping off and root diseases with conservation tillage, use of fungicide seed treatments for soybeans might be expected to increase with conservation tillage.

REFERENCES

Banks, P. A. and E. L. Robinson. 1983. Activity of acetochlor, alachlor, and metolachlor as affected by straw. Proc. South. Weed Sci. Soc. 36:394.

Becker, R. L. 1978. Effects of tillage on patterns of annual weed seed germination and emergence. M.S. Thesis, Iowa State University. 74 pp.

Becker, R. L. 1982. Perennial weed response to soil tillage. Ph.D. Thesis, Iowa State University. 120 pp.

Boosalis, M. G., B. Doupnik, and G. N. Odvody. 1981. Conservation tillage in relation to plant disease. In Handbook of Pest Management in Agriculture, Vol. 1, CRC Press, Boca Raton, Fla. 600 pp.

Boosalis, M. G., D. R. Sumner, and A. S. Rao. 1967. Overwintering conidia of *Helminthosporium turcicum* on corn residue and in soil in Nebraska. Phytopathology 57: 990-996.

Brooks, D. H. and M. G. Dawson. 1968. Influence of direct-drilling of winter wheat on incidence of take-all and eyespot. Ann. Appl. Biol. 61:57-64.

Burns, E. E. and M. C. Shurtleff. 1973. Observations of *Physodermata maydis* in Illinois: Effects of tillage practices in field corn. Plant Dis. Reptr. 57:630-633.

Busching, M. K. and F. T. Turpin. 1976. Oviposition preferences of the black cutworm moths among various crop plants, weeds, and plant debris. J. Econ. Entomol. 69(5):587-590.

Christensen, R. G., P. Wise, J. E. Lake, B. A. Julian, T. Schach, and J. B. Morrison. 1985. Lake Erie conservation tillage demonstration projects: evaluating management of pesticides, fertilizer, residue to improve water quality. USEPA publication. 22 pp.

Colvin, T. S., R. M. Cruse, D. R. Timmons, A. Musselman, H. J. Brown, and M. Culik. 1985. Southeast Iowa conservation tillage research project Middleton, Iowa. Annual progress report. Iowa State University. 59 pp.

Cook, R. J., M. G. Boosalis, and B. Doupnik. 1978. Influence of crop residues on plant diseases. p. 147-163. In Crop Residue Management Systems. Am. Soc. Agron. Spec. Pub. 31. Madison, Wis.

Cook, R. J. and J. T. Waldher. 1977. Influence of stubble-mulch residue management on Cercosporella foot rot and yields of winter wheat. Plant Dis. Reptr. 61:96-100.

Cox, R. W., F. C. Collins, and J. P. Jones. 1976. Soybean seedling diseases associated with double cropping. Ark. Farm Res. 25:(3)5.

Crutchfield, D. A., G. A. Wicks and O. C. Burnside. 1986. Effect of winter wheat (Triticum aestivum) straw mulch level on weed control. Weed Sci. 34:110-114.

Daft, G. C. and C. Leben. 1973. Bacterial blight on soybeans: field-overwintered Pseudomonas glycinea as possible primary innoculum. Plant Dis. Rept. 57:156-157.

De la Cruz, R. 1974. Weed seedling emergence depths under field conditions. Ph.D. Thesis. Iowa State University. 115 pp.

Delvo, H. W. 1984. Inputs outlook and situation. Economic Research Service Report IOS-6. p. 4.

Doupnik, B. Jr., and M. G. Bousalis. 1980. Ecofallow - a reduced tillage system-and plant diseases. Plant. Dis. 1:31-35.

Duffy, M. 1983. Pesticide use and practices. 1982. Agricultural Information Bulletin No. 462, Economic Research Service. 14 pp.

Elliot, E. S. 1976. Diseases of Forage Crops. United States Department of Agriculture/Current Research Information Service Progress Report.

Erbach, D. C. and W. G. Lovely. 1975. Effect of plant residue on herbicide performance in no-tillage corn. Weed Sci. 23:512-515.

Fawcett, R. S. 1985. Weed control in conservation tillage. Cooperative Extension Service, Iowa State University. 12 pp.

Fawcett, R. S. and M. D. K. Owen. 1984. Influence of nozzle type and carrier volume on residual herbicide activity in high residue no-till corn production. North Cent. Weed Cont. Conf. Res. Rpt. 41:130-131.

Fawcett, R. S., M. D. K. Owen and P. C. Kassel. 1983. Early preplant treatments for weed control in no-till corn and soybeans. Proc. North Cent. Weed Cont. Conf. 38:112-117.

Ferrant, N. P. and R. B. Carroll. 1979. Fusarium wilt of soybean and effect of tillage practices on occurrence of Fusarium species in roots and soil. Phytopathology 69:534-535.

Fleige, H. and K. Baeumer. 1974. Effects of no-tillage on organic carbon and total nitrogen content, and their distribution in different N fractions in Loessial soils. Agr. Ecosystems 1:19-29.

Foster, D. E. 1986. Insect management in reduced tillage corn. Cooperative Extension Service, Iowa State University. 2 pp.

Ghadiri, H., P. J. Shea, and G. A. Wicks. 1984. Interception and retention of atrazine by wheat (Triticum aestivum L.) stubble. Weed Sci. 32:24-27.

Gray, M. E. 1986. The influence of conservation tillage practices on corn rootworms. Proc. Iowa Ferilizer and Ag Chemical Dealers Conference 38:CE 2158c.

Gregory, W. W. and G. J. Musick. 1976. Insect management in reduced tillage systems. Bull. Entomol. Soc. Am. 22(2):302-304.

Griffith, D. R., J. V. Mannering, and W. C. Moldenhauer. 1977. Conservation tillage in the eastern corn belt. J. Soil Water Conserv. 32:20-28.

Hanthorn, M. and M. Duffy. 1983. Corn and soybean pest management practices for alternative tillage strategies. Inputs Outlook and Situation. USDA/ERS Publication IOS-2. p. 14-17.

Hilty, J. W., C. H. Hadder, and F. T. Garden. 1979. Response of maize hybrids and inbred lines to gray leaf spot disease and the effects on yield in Tennessee. Plant Dis. Reptr. 63:515-518.

Hinkle, M. K. 1983. Problems with conservation tillage. J. Soil and Water Cons. 38:201-206

Holm, R. E. 1972. Volatile metabolites controlling germination in buried weed seeds. Plant Phys. 50:293-297.

Hood, A. E. M. 1965. Plowless farming using "Gramoxone". Outlook Agric. 4(6):286-294.

Kells, J. J., C. E. Rieck, R. L. Blevins, and W. M. Muir. 1980. Atrazine dissipation as affected by surface pH and tillage. Weed Sci. 28:101-104.

LaCroix, L. J. and D. W. Staniforth. 1964. Seed dormancy in velvetleaf. Weeds 12:171-174.

Lugo, R. V. 1984. Germination of weed seeds in response to tillage and soil moisture variables. M.S. Thesis, Iowa State University. 119 pp.

Martin, C. D., J. L. Baker, D. C. Erbach, and H. P. Johnson. 1978. Washoff of herbicides applied to corn residue. Trans. Amer. Soc. Agric. Eng. 21:1, 164-168.

Musick, G. J. and D. L. Collins. 1971. Northern corn rootworm affected by tillage. Ohio Rep. 56:88-91.

Musick, G. J. and H. B. Petty. 1973. Insect control in conservation tillage systems. p. 120-125 In Conservation tillage: The proceedings of a national conference. Soil Conserv. Soc. Am., Ankeny, Iowa.

Musick, G. J. and H. B. Petty. 1974. Insect control in conservation tillage systems. Conservation Tillage, A Handbook for Farmers. Soil Cons. Soc. Am. 52 pp.

Musick, G. J. and L. E. Beasley. 1978. Effect of the crop residue management system on pest problems in field corn (Zea mays L.) production. p. 173-187 In Crop Residue Management Systems, Amer. Soc. Agron., Madison, Wis.

Pareja, M. R. and D. W. Staniforth. 1985. Seed-soil microsite characteristics in relation to weed seed germination. Weed Sci. 33:190-195.

Pareja, M. R., D. W. Staniforth, and G. P. Pareja. 1985. Distribution of weed seed among soil structural units. Weed Sci. 33:182-189.

Phillips, S. H. 1984. Other pests in no-tillage and their control. p. 171-189 In R.E. Phillips and S. H. Phillips (ed), No Tillage Agriculture. Van Nostrand Reinhold Co., New York.

Phillips, R. E., R. L. Blevins, G. W. Thomas, W. W. Frye, and S.H. Phillips. 1980. No-tillage agriculture. Science 208:1108-1113.

Roane, C. W., R. L. Harrison, and C. F. Gentner. 1974. Observation on gray leaf spot of maize in Virginia. Plant Dis. Reptr. 58:456-459.

Schaefer, K. P. 1984. The influence of temperature and seed bed tillage on the emergence and early growth of corn and two weed species, giant foxtail and velvetleaf. Ph.D. Thesis, Iowa State University. 96 pp.

Schnappinger, M. G., C. P. Trapp, J. M. Boyd, and S. W. Pruss. 1977. Soil pH and triazine activity in no-tillage corn as affected by nitrogen and lime applications. Northeastern Weed Sci. Soc. Proc. 31:116.

Schuster, M. L. 1975. Leaf freckles and wilt of corn incited by *Corynebacterium nebraskense* Schuster, Hoff. Mandel, Lazar. 1972. Nebr. Agric. Exp. Stn. Res. Bull. 270. 40 pp.

Slack, C. H., R. L. Blevins, and C. E. Rieck. 1978. Effect of soil pH and tillage on persistence of simazine. Weed Sci. 26:145-148.

Steinsiek, J. W., L. R. Oliver, and F. C. Collins. 1982. Allelopathy potential of wheat (*Triticum aestivum*) straw on selected weed species. Weed Sci. 30:495-497.

Taylorson, R. B. 1970. Changes in dormancy and viability of weed seeds in soils. Weed Sci. 18:265-269.

Taylorson, R. B. 1972. Phytochrome changes in dormancy and germination of buried weed seeds. Weed Sci. 20:417-422.

Taylorson, R. B. and S. B. Hendricks. 1972. Interactions of light and a temperature shift on seed germination. Plant Phys. 49:127-130.

Thilsted, E. and D. S. Murray. 1980. Effect of wheat straw on weed control in no-till soybeans. Proc. South. Weed Sci. Soc. 33:42.

Thomas, S. H. 1978. Population densities of nematodes under seven tillage regimes. J. Nematol. 10:24-27.

Triplett, G. B., Jr., and D. M. Van Doren, Jr. 1969. Nitrogen, phosphorus and potassium fertilization of no-tilled maize. Agron. J. 61:637-639.

Triplett, G. B., Jr. and G. D. Lytle. 1972. Control and ecology of weeds in continuous corn grown without tillage. Weed Sci. 20:453-457.

Wesson, G. and P. F. Wareing. 1969. The role of light in the germination of naturally occurring populations of buried weed seeds. J. Exp. Bot. 20:402-413.

White, D. G., J. Yanney, and T. A. Natti. 1979. Anthracnose stalk rot. Proc. Ann. Corn Soybean Res. Conf. 34:1-16.

Wicks, G. A. 1985. Weed control in conservation tillage systems-small grains. p. 77-91 In A. F. Wiese, ed. Weed control in limited-tillage systems. Weed Sci. Soc. Amer., Urbana, Ill.

Wicks, G. A. and B. R. Sommerhalder. 1971. Effect of seedbed preparation for corn on distribution of weed seed. Weed Sci. 19:666-668.

Wicks, G. A. and R. N. Klein. 1985. Ecofarming - an integrated crop protection system. p. 72-80 In Integrated Pest Management into Conservation Tillage Proceedings of North Central Region Workshop, St. Louis, Missouri.

Williams, J. L., Jr. and G. A. Wicks. 1978. Weed control problems associated with crop residue systems. Crop Residue Management Systems. Amer. Soc. Agron. Spec. Pub. 31. Chapter 9. pp. 165-172.

CHAPTER 3

OVERVIEW OF NITROGEN MANAGEMENT FOR CONSERVATION TILLAGE SYSTEMS: AN OVERVIEW

G. W. Randall,
University of Minnesota, Waseca, Minnesota

V. A. Bandel,
University of Maryland, College Park, Maryland

INTRODUCTION

Most fertilizer research over the last 30 years has been conducted with conventional tillage. Usually conventional tillage has implied primary tillage with a moldboard plow followed by various secondary tillage operations including disking, field cultivating and harrowing. Consequently, nitrogen (N) recommendations have been based on a crop management system that is much different that the conservation tillage systems now gaining popularity.

Conservation tillage (CT) strictly defined by the USDA-SCS describes a field situation where 30% of the soil surface after planting is covered with plant residues from the previous crop(s). Residue coverage can range from somewhat less than this (depending on tillage system used and previous crop) to almost complete coverage. These residue conditions coupled with the fact the there is less physical disturbance and inversion of the soil plow layer requires a whole new fertilizer research program. This program must identify both the best management practices (BMP) to maximize economic return and minimize enviromental impact and the climatic/soil/crop interactions which effect the BMP's.

Nitrogen is capable of undergoing a number of transformations in what is commonly called the "N cycle" (Stevenson, 1982). Additional information on these

Effects of Conservation Tillage on Groundwater Quality: Nitrates and Pesticides, Terry J. Logan et al., eds.

transformations and the use of N fertilizers has been summarized recently by Boswell et al., (1985). Some of these transformations, e.g. immobilization, denitrification, and volatilization, can be directly affected by surface residues; hence, the tillage system used. (A more specific and thorough description of how CT affects these transformation will be given in lead and response papers in this workshop by Gilliam and Schepers, respectively.)

N MANAGEMENT FOR VARIOUS CT SYSTEMS

Conservation tillage systems imply less rigorous tillage activity with the end result being an accumulation of plant residues from the previous crop(s) remaining on the soil surface after planting. There are a number of different CT systems now in place in the U.S. Moreover, numberous types and kinds of implements are available to establish these systems. For the purposes of clarity and brevity we can divide these CT systems into two broad categories: full-width versus strip tillage. (A thorough description of CT systems will be given in this workshop by Mannering and Schertz.)

In the full-width systems, primary and/or secondary tillage is accomplished across the full width of the tillage or planter equipment. These systems include the chisel plow, disk, field cultivator, powered rotary tiller and in some cases the subsoiler (depth of operation often between 25 and 45 cm). The end product of these operations is substantial soil disturbance, especially of the soil surface, with varying degrees of soil inversion and residue incorporation uniformly distributed across the field.

Strip tillage is defined as performing a limited amount of tillage in a narrow strip centered on the planted row. These systems prepare a seedbed within the row to improve seed:soil contact while leaving the inter-row area untilled. Strip width varies from as narrow as two offset coulters penetrating and forming a small slit (commonly called no tillage), to wider strips (< 5 cm) where the area is tilled with fluted or ripple coulters, to strips that may be 15-30 cm wide. These wider strips include the ridge-plant system and other rotary and till-plant systems that clear the row area of residue without disturbing the inter-row areas. In-row subsoilers used on the compacted soils of the Southeastern U.S. fit into this category.

Chisel Plow

The chisel plow tillage system is rapidly becoming popular and in some areas has replaced the moldboard plow as the conventional tillage system. Chisel plows equipped with coulters or disks to cut the residue are capable of primary tillage even where residue levels are very high. They leave the surface quite rough while incorporating from 25 to 50% of the residue. Those with twisted, wide (8-10 cm) shovels incorporate more residue but leave the soil surface in a rougher condition. The surface roughness traps moisture, increases infiltration, and reduces erosion. Surface-applied fertilizers are incorporated to a 10-15 cm depth when chisels are operated customarily at a 18-23 cm depth.

A secondary tillage operation usually consisting of a disking and/or field cultivation is necessry to smooth the soil for planting and can satisfactorily incorporate surface-applied N. This operation further incorporates residue so that surface coverage often is about 35% when following corn but is reduced to less than 10% when following soybeans or other low residue crops. With anywhere from 20 to 40% surface residue coverage, urea-containing N fertilizers broadcast-applied after secondary tillage are highly susceptible to NH_3 volatilization losses. Therefore, injection of anhydrous NH_3 (AA) and UAN solutions or ammonium nitrate (AN) broadcast on the soil surface are usually preferred N management programs. Differences among N sources or placement positions are negligible when following low residue crops e.g., soybeans, unless the soils are highly calcareous. Under these high soil pH conditions (pH>7.4), incorporation of N is recommended.

Subsoilers operated at depths >25 cm with the intention of ameliorating compaction layers can be thought of as similar to traditional chisel plows with respect to CT. Surface roughness, residue incorporation, and the need for secondary tillage provide the same N management conditions.

Disk

The disk has long been used as a secondary tillage implement for preparing the seedbed just prior to planting and for herbicide incorporation. Most of these disks have 40-50 cm blades and are rather light. Therefore, when used as a primary tillage tool they seldom penetrate deeper than 10 cm and incorporate from 30 to 70% of the residue, depending on the amount and type of residue. Even though sufficient residue may have been left on the surface, the surface is left quite smooth and is susceptible to water runoff and wind erosion. Incorporation of surface-applied fertilizers is seldom much below 5-8 cm.

Both "one-way" and "heavy" disks are also available for primary tillage. These implements have disk blades up to 90 cm in diameter and thus can easily penetrate to a 15-20 cm depth. Horsepower requirements are high and unfortunately the amount of residue left on the surface is often less than desired. Also, compaction with these heavy disks is thought to be a problem when soils are relatively wet.

Depending on the type of disk used, the amount of residue left on the surface is usually similar to that of the chisel system. Therefore, N management alternatives and recommendations are the same.

Rotary-Till

Powered rotary tillers can be used to prepare the seedbed while incorporating fertilizers and pesticides. Planting units usually are attached to the rotary tiller, making tillage and planting a one-pass operation. Well suited to medium-textured soils, the rotary-till system can prepare a finely pulverized seedbed, providing excellent seed-to-soil contact for germination. However, depending on use, the surface may be residue-free after planting and subject to erosion and crusting after rainfall (Dickey et al., 1986). With full-width rotary tillage, surface N application just prior to the tillage is acceptable. Nitrogen applications after planting will be most efficient if they are injected. Surface applications of urea-containing sources may be susceptible to volatilization losses if surface residue amounts are >10% or may be lost with the runoff water and sediment if erosion occurs.

Rotary-till implements can also be used in strip tillage systems. Shallow tillage of strips 25-38 cm wide can give an excellent seed bed while providing additional erosion control by leaving more residue between the rows. Under these conditions N management becomes more critical. Injection of N either prior to or after planting should result in optimum N efficiency.

Ridge Plant

An alternative to no tillage in the North Central states is the ridge-plant system. This is a one-pass system involving a slight amount of tillage by the planter as it places the seed into a preformed ridge area (Randall, 1984). The 10-18 cm ridges are formed during cultivation of the previous crop. Sweeps or other ridge-cleaning devices mounted on the planter scrape the top 1-5 cm of soil and residue off the ridge top into the row middles. This allows for a residue-free, warm seed zone with ample amounts of residue between the rows for erosion control. Surface residue coverage across the rows will range from 10% to 40% depending on the previous crop and the amount of soil removed from the ridge. Removing over 5 cm of the ridge covers much of the residue between the rows and defeats the purpose of the system.

Because surface residue amounts are highly concentrated between the ridges, yet may average up to 40% coverage, a unique opportunity exists for N management. Nitrogen inefficiency from surface-applied N can be expected due to volatilization losses and immobilization. In a Minnesota study, Randall and Langer (1982) applied three N sources (UAN, urea and AA) at three different times (preplant, emergence, and 8-leaf stage) to corn (Table 1). They found poorest N efficiency with UAN and AA applied at the emergence stage. The UAN was apparently volatilized or immobilized by the residue (23% coverage). On the other hand, AA escaped from the soil at the time of injection due to sealing problems caused by the large amount of residue that was incorporated between the ridges by the planter. This residue was not well decomposed at this stage; therefore, the slit caused by the applicator knife did not seal well and NH_3 vapors escaped. Application of N immediately prior to planting and incorporation of the surface-applied materials by the planter resulted in high yields and no difference among N sources. Highest yields and greatest N efficiency were obtained with the sidedress application at the 8-leaf stage. In subsequent years when dry conditions followed application, yields were reduced with the sidedressed UAN and urea treatments (data not shown). These treatments were incorporated to a 2.5-5 cm depth with a cultivator, but apparently the N remained positionally unavailable near the soil surface. The AA, which was injected to a 18 cm depth, gave highest yields.

Table 1. Effect of N source and time/method of application on corn yields with a ridge-plant system in 1981 (Randall and Langer, 1982).

	Application time/method		
N source	Preplant (broadcast)[1]	Emergence (broadcast)[2]	8-leaf (sidedress)[3]
UAN (28%)	10.42	9.60	10.61
Urea	10.55	10.48	10.42
AA[4]	10.30	9.92	10.55

1. UAN and urea were applied preplant and incorporated with the planter.
2. UAN and urea were broadcast applied and not incorporated.
3. UAN and urea were sidedressed in a band near the row and cultivated in.
4. Knifed-in between 76 cm rows.

Yield results obtained in these studies with ridge tillage indicate the importance of N placement associated with the source of N and time of application. Regardless of tillage system, N that is positionally unavailable to the growing crop remains in the soil profile as NO_3 after harvest of the crop and is highly susceptible to leaching through the soil profile and into groundwater.

No Tillage

In no tillage (NT) systems the seed is place in a narrow slit made by a disc-opener. Residue disturbance is minimal. Thus, depending on the previous crop and the climate, residue coverage may easily be over 90%. Usually the residue is uniformly spread across the soil surface but can be more concentrated behind the combine if the straw spreaders do not distribute the residue evenly.

Management of N fertilizers is critical and must be conducted carefully if maximum N efficiency is to be realized. A thorough review of N management in the NT system was reported by Wells (1984). Some N management problems have resulted from the fertilizer industry's trend away from AN and toward urea and urea-based sources. Indications are that these N sources often present agronomic and subsequent economic problems for the farmer (Bandel et al., 1980; Touchton and Hargrove, 1982). It is well known that, under favorable conditions, significant quantities of N can be lost to the atmosphere from surface-applied urea-containing fertilizers due to ammonia volatilization. Consequently, yeilds are often reduced (Bandel et al., 1980; Fox and Hoffman, 1981; Touchton and Hargrove, 1982). In some cases, ammonia losses from surface-applied urea may be reduced by appropriate use of a nitrification inhibitor (Frye et al., 1981).

Nitrogen efficiency can be improved significantly by proper fertilizer placement or if sufficient rainfall occurs at a fortuitous time. Fox and Hoffman (1981) reported insignificant ammonia volatilization losses from unicorporated urea if at least 1 cm of rain fell within 3 days after the urea was applied, but if no rain fell within 6 days, the losses could be over 30%. University of Maryland research (Bandel, 1984) showed injected and dribbled UAN to give statistically higher yields compared to broadcast UAN at all locations (Table 2). Although these treatments would not accommodate simultaneous tank-mix herbicide applications, they would provide greater N efficiency and subsequently higher economical profits.

Further Maryland research (Bandel, 1986) conducted over 6 years at three locations (18 location-years) show a consistent advantage for the injection of UAN (Table 3). The broadcast UAN treatment in May (within 2 to 3 days of planting) was considered to be a standard treatment similar to that practiced by many farmers who tank-mix their pesticides and broadcast-apply with UAN. Yields were increased 7% by injecting at planting compared to broadcasting. Broadcast application 4 weeks after planting increased yields 10% compared to the earlier application. However, by delaying the application and by injecting instead of broadcasting, yields were increased. Dribble applications over the last 9 location-years did not produce yields as high as when the UAN was injected.

Table 2. Influence of N source and placement on no-tillage corn grain yields in Maryland in 1982 (Bandel, 1984).

Poplar[2] N Treatment	Location			
	Wye[1]	Poplar[1] Hill	Poplar[1] Hill	Hill
	Mg/ha			
CHECK	2.07	1.95	2.65	2.64
Ammonium Nitrate	7.03	9.73	8.92	10.30
UAN, Broadcast	6.22	7.54	8.54	9.99
UAN, Dribbled	7.54	9.86	9.36	11.05
UAN, Injected	7.79	10.49	9.80	11.18

1. N rate = 134 kg/ha
2. N rate = 179 kg/ha

Table 3. Relative influence of N placement and time of application on average no-till corn grain yields (Bandel, 1986).

UAN Placement[1]	1980-85 Avg		1983-85 Avg	
	May[2]	June[3]	May[2]	June[3]
	relative yield			
Broadcast	100	110	100	112
Dribble	---	---	105	113
Inject	107	114	120	121

1. N rate = 134 kg/ha
2. Applied within 2 to 3 days of planting
3. Applied 4 weeks after planting

In addition to the accumulation of residues on the soil surface of NT, there is some evidence that a greater number of macropores exists near the surface compared to tillage systems that disturb the soil. In wet soils these large pores can rapidly transport large quantities of water, potentially containing nitrates, down through the soil profile (Horton, 1986). In Kentucky leaching of NO_3 during the growing season has been shown to be greater with NT than with conventioanl tillage (Thomas et al., 1973). This may have been due to less evaporation, less runoff and more large pores in the undisturbed soil. After 10 years of continuous tillage for corn in Minnesota, Randall et al. (1980) found NO_3-N amounts in the 0-3 m profile of the NT and chisel plow treatments to be 50 and 74% of that found in the moldboard plow treatment (Table 4). These data suggest that more of the N may have been leached or denitrified from the NT treatment and/or that less soil N was mineralized with NT.

Studies to ascertain the effects of tillage on nitrate losses to tile lines were conducted recently on a Webster clay loam in Minnesota (Randall and Kelly, 1986). They found very little difference in tile water flow, NO_3-N concentrations or NO_3 flux between the moldboard plow and NT systems after 4 years (Table 5). However, more detailed measurements by Culley (1986) at this site showed a greater proportion of macropores and greater downward movement of a bromide tracer added with water on the non-wheeltracked NT plots. Contradictory results were found on a similar soil by Lindstrom et al. (1981). After 10 years of continuous tillage they concluded that infiltration was actually lower with NT than with chisel or moldboard plow tillage due to a consolidated soil surface with high bulk density.

Table 4. Nitrate-N distribution in the 0-1.5 m and 0-3 m profile of a Webster clay loam after 10 years of continuous tillage (Randall et al., 1980).

Profile depth	Tillage System		
	Moldboard plow	Chisel plow	No tillage
	kg NO_3-N/ha		
0 - 1.5	264	186	120
0 - 3	614	452	306

Table 5. Cumulative effects of two tillage systems over 4 years (Randall and Kelly, 1986).

Parameter	Tillage Systems	
	Mb. plow	No tillage
Corn grain removed (Mg/ha)	33.3	31.2
N removed in grain (kg/ha)	388	353
Tile flow (ha/cm)	104	112
NO_3-N lost in tile (kg/ha)	99	110

Fertilizer Injection

A dilemma exists currently between the desire to incorporate chemicals to reduce runoff and increase efficiency and the desire to leave as much crop residue on the soil surface as possible to protect against erosion (Baker and Laflen, 1983). When residue amounts are high or slopes gentle, some tillage to incorporate the fertilizer may be acceptable. But if residue amounts are low, e.g., soybeans, or slopes steep, this is usually not desirable.

For this reason, new fertilizer application methods need to be developed and researched. Currently, there is interest in the air-blast (Nutriblast), coulter-stream, and point injector systems. The air-blast system requires pressures as high as 141 kg/cm^2 (2000 psi) or more to physically force the fertilizer solution through the residue and into the top few centimeters of soil. The coulter-stream applicator cuts through the residue with the coulter and either sprays or dribbles the solution in the slit formed below the residue by the coulter. The point injector applies fertilizer through a rolling spoked wheel about 10 cm into the soil. Essentially no soil or residue disturbance was reported when the applicator was used in NT continuous corn (Baker et al., 1985; Dawelbeit et al., 1981). These systems appear to have promise in placing the N fertilizer below the surface residue but are limited to fertilizer solutions. Narrow profile knives appear to be best for AA but some disturbance will occur. Application of dry N sources below the surface is limited because wider knives are essential which cause greater soil disturbance.

In summary, surface residue accumulations will increase the potential for greater immobilization of broadcast fertilizer N, will create conditions conducive for NH_3 volatilization losses of surface-applied ammoniacal N sources, and will also provide a condition for greater potential denitrification losses (Randall, 1984). These potential losses are often most severe with the urea-containing N sources, e.g., urea and urea-ammonium nitrate (UAN) solutions. For these reasons and based on recent research (Bandel et al., 1980; Griffith, 1974; Mengel et al., 1982) most agronomists suggest a carefull selection of the N source or the injection of N materials below the zone of residue accumulation when making fertilizer N recommendations to farmers using CT systems. To maximize efficiency and profit and minimize environmental effects, improved N management and application techniques will be necessary (Randall et al., 1985).

CURRENT STATE AND REGIONAL RECOMMENDATIONS

Letters were sent to an extension agronomist in each of the 48 contiguous states asking for their comments and state's recommendations regarding N management with CT. Each agronomist was asked to comment specifically on the three following questions and to provide data if possible.

1) Discuss your state N recommendations with special attention to any differences in N management techniques that may be recommended for conventional tillage, no-tillage, and other conservation tillage methods. Consider N rate, source, time of application, placement, etc.
2) Do you believe that N is utilized just as efficiently, more efficiently, or less efficiently under conservation tillage as compared to conventional tillage?
3) Do you believe that conservation tillage "threatens" the environment, particularly surface and groundwater, more, less, or the same as conventional tillage?

Based on the excellent cooperation of these colleagues we have attemted to paraphrase and assemble their thoughts and comments into Table 6. We apologize if some of their remarks have been taken out of context or if misinterpretations have resulted.

In general, there was very little difference in N recommendations for various tillage systems in the Northeast, Middle Atlantic, and Southeastern states. Most states do not currently recommend additional N for no-till. But, where an extra amount is suggested, the justification for the increase is based upon the reduced availability of soil N and the increased crop demand due to higher yields. All state agronomists that responded agreed upon the importance of careful N management on NT corn.

In the North Central states, where annual rainfall generally exceeds 70 cm and corn is the dominant crop, N recommendations are raised slightly with CT in three states. These increased N recommendations are usually targeted toward NT. In addition, almost all states recommend the incorporation of urea-containing fertilizers because of potential volatilization and/or immobilization losses. Proper placement is stressed more than N rate or N source when surface residues exist. Anhydrous ammonia is generally the preferred source for greatest efficiency with urea being the least preferred. Recommendations for 10 to 30% more N are sometimes made when surface applying urea-based fertilizers to residue-covered soil without incorporation.

Although N rates are not changed for tillage systems in Louisiana, the same caution toward surface application of N exists under their conditions.

Under the humid conditions of the Central states, there was fairly universal feeling that fertilizer N was used as efficiently with CT systems as with conventional tillage. Data are limited, however. Some data do exist, though, that indicate reduced mineralization of soil OM and thus the need for more fertilizer N under CT. Because data are also limited on the role of CT with respect to N contamination of ground and surface waters, little speculation or discussion was generated among colleagues. Most thought little difference would exist in potential groundwater contamination between tillage systems.

Table 6. State recommendations for N management with conservation tillage.

Region/State	Are N rate, timing, and placement recommendations adjusted for CT?	Is N used more efficiently with CT?	Does CT "threaten" ground or surface water with respect to N more than Conv. Till.?	Other comments
Northeast, Middle Atlantic, and Southeast	Rates generally not adjusted. UAN should be either injected or dribbled.	-----	-----	1) Ideally, most N should be applied about 4 to 6 weeks after planting. 2) Some states suggest on the average 30 to 45 kg N/ha more N for CT but varies from 0 to 60 depending on soil and previous crop. Usually justified with higher yield.
North Central				
Illinois	Increase rate 10 to 20% w/NT, mostly because urea or UAN are commonly used.	From 10 to 20% more N is needed w/NT perhaps due to less mineralization or greater denitrification.	-----	1) Discourage the use of urea-containing products when they can't be incorporated soon after application.
Indiana	Rates are increased only when surface applied to compensate for volatilization and/or immobilization.	Similar if volatilization and/or immobilization are minimized.	Perhaps more so with surface application. The immobilized N can be mineralized later in season & then is susceptible to leaching. Also, contiguous macropores with NT may accelerate leaching losses of nitrate.	1) Stress placement of N below the soil surface. This also removes the weather factor. 2) Effectiveness of surface-applied N is inversely proportional to the amount of surface residues. 3) AA & UAN injected are preferred, urea is least preferable.

Table 6. State recommendations for N management with conservation tillage. Cont.

Region/State	Are N rate, timing, and placement recommendations adjusted for CT?	Is N used more efficiently with CT?	Does CT "threaten" ground or surface water with respect to N more than Conv. Till.?	Other comments
Indiana (Continued)				4) N rate is increased 15 to 30% in weed and feed applications to soils with surface residues.
Iowa	No, but little data available.	No data to indicate more or less.	About the same.	
Michigan	No rate recommendation adjustments unless when surface-applied to high residue conditions. Little data available.	Little data but probably as efficient.	About the same.	1) Potential sizable losses of urea or UAN when surface-applied to residue. Ammonium nitrate is best under these conditions. 2) Immobilization of surface-applied N may require 10 to 20% more N. 3) Surface-applied N accentuates soil acidity.
Minnesota	N rates increased by 45 kg N/ha for NT when corn follows corn or a small grain. N rates increased by 22 kg N/ha for ridge tillage when corn follows corn or a small grain.	About the same if placed properly.	About the same, but little data available.	1) Place sidedressed N below the soil surface. 2) Incorporate urea containing N sources within 2 to 3 days after application.

Table 6. State recommendations for N management with conservation tillage. Cont.

Region/State	Are N rate, timing, and placement recommendations adjusted for CT?	Is N used more efficiently with CT?	Does CT "threaten" ground or surface water with respect to N more than Conv. Till.?	Other comments
Minnesota (Continued)				3) When following alfalfa or clover with NT corn, increase N rates by 67 kg N/ha if poor stand or by 45 kg N/ha if good stand of previous crop.
Ohio	No rate adjustments. Band or inject UAN.	Probably as efficient if applied properly and in correct amounts.	No, most losses are due to denitrification, immobilization and volatilization. Little is left for leaching.	1) Average application rate is too high regardless of tillage for the yields obtained. 2) AA and AN are preferred w/NT, broadcast urea and UAN should be avoided unless 1 cm rain immediately after application. 3) Split applications sometimes helpful. 4) Surface applied N depresses soil pH.
Wisconsin	Increase N rate for corn by 34 kg N/ha when a tillage system has 50% surface residue coverage. Recommend injection if at all possible.	Less because greater opportunities for immobilization, leaching and denitrification. Studies show less soil N available to NT corn.	CT does not generally increase the risk of N contamination.	1) Because of volatilization losses of surface-applied urea-containing N sources, either inject the N or apply a 15 to 20% higher rate.

Table 6. State recommendations for N management with conservation tillage. Cont.

Region/State	Are N rate, timing, and placement recommendations adjusted for CT?	Is N used more efficiently with CT?	Does CT "threaten" ground or surface water with respect to N more than Conv. Till.?	Other comments
South Central				
Louisiana	No	The same or slightly less efficiently because volatilization & immobilization of surface applications.	About the same, however increased infiltration implies a greater potential for leaching to groundwater.	1) When surface applying urea, farmers may need to apply 10 to 15% more N.
Great Plains				
Colorado	No	No data to indicate more or less.	CT improves surface water with no difference for groundwater.	
Kansas	No	No data to indicate more or less if proper placement. However, under dry conditions, N may be utilized more efficiently.	Less, because CT can give better moisture utilization and hence higher fertilizer uptake.	1) UAN should not be put on residue unless it is incorporated.
Montana	Recommended N rate based on yield goal (soil moisture plus anticipated precipitation) plus residual soil NO_3. Increase recommendation by 20 kg N/metric ton of surface residue.	More	Less	1) Apply urea containing fertilizers to the side or below the seed seed. 2) Additional N required for recrop fields compared to fallowed fields.

Table 6. State recommendations for N management with conservation tillage. Cont.

Region/State	Are N rate, timing, and placement recommendations adjusted for CT?	Is N used more efficiently with CT?	Does CT "threaten" ground or surface water with respect to N more than Conv. Till.?	Other comments
Nebraska	No	Probably more especially under dryland conditions.	About the same.	1) Greatest N efficiency with incorporated N, especially with urea containing fertilizers applied to high pH soils or those with surface residues. 2) Urea is not recommended for surface application w/o incorporation at pH >7.0. 3) Sidedress may be superior to preplant application with CT systems as long as sufficient moisture is available in the zone of fertilizer application.
North Dakota	No adjustment in rate. All N sources should be placed below surface residues.	Yes, by improving water efficiency.	----	
Oklahoma	Increase N rate 20 kg N/ha for each metric ton of residue on the surface when planting wheat or grain sorghum.	No, more N is immobilized. Also data show that more N is needed w/CT to get equal yields.	Less	

Table 6. State recommendations for N management with conservation tillage. Cont.

Region/State	Are N rate, timing, and placement recommendations adjusted for CT?	Is N used more efficiently with CT?	Does CT "threaten" ground or surface water with respect to N more than Conv. Till.?	Other comments
South Dakota	No adjustment in rates. Apply N below residue accumulation to prevent immobilization and/or volatilization.	About the same.	Less	
West				
Oregon	No	Little data yet but probably equal.	No difference	1) Band apply N for both systems. 2) CT limits runoff, which causes greater percolation that may leach N, but crop yields & N uptake also higher.

Under the drier conditions in the Great Plains where wheat and sorghum dominate, N rates are seldomly changed with CT. In Montana and Oklahoma the N rates are increased 20 kg/ha for each metric ton of residue on the surface. This recommendation is made to account for greater immobilization of the N by the residues, reduced mineralization of the soil OM, as well as a greater yield potential due to moisture conservation and utilization. Similar to the Central states some recommend incorporation of fertizilier N applied to high pH soils and/or those convered with residue. Because of better moisture utilization, N efficiency with CT systems was thought to be better than with conventional systems by a majority of the states. Most also indicated that CT systems "threatened" ground and surface water less than conventional tillage.

USAGE AND TRENDS IN NITROGEN SOURCES AND APPLICATION RATES

Total N fertilizer usage has increased dramatically in the last 30 years. In the 20 years between 1955 and 1975, N usage in the U.S. rose by over 6.0 million metric tons to a total of 7.8 million metric tons (Table 7). The fastest rate of increase occurred in the 5-year period between 1975 and 1980 when another 2.5 million metric tons was used. In the 1974-84 period N usage increased by 2% per year. Since 1980, N usage has leveled off at slightly over 10 million metric tons.

This change in N usage is closely coupled with increased production of corn and wheat (Table 8). Corn acreage increased 23% in the 20-year period from 1965 to 1984 while yields increased 44%. This resulted in a total production increase of 87%. Wheat production increased even more dramatically over this period. Acreage increased by 38% and yields by 46%, resulting in total production being increased by 97%.

As might be expected N usage is highest in the North-Central and West South-Central states where large acreages of corn and wheat predominate (Table 7). These also happen to be the areas where a number of CT systems are being implemented in an effort to reduce erosion and improve economic return. The 12% decline in the amount of N used in the West North-Central states in 1984 may be an after-effect of the PIK program in 1983. Lower amounts of N were often recommended in 1984 because corn acreage was reduced substantially in 1983 with subsequent planting of cover crops. These were often plowed down in the late summer and provided a short fallow season.

Table 7. Total fertilizer nitrogen applied to agricultural crops from 1955 through 1984 (Hargett & Berry, 1985).

Area	Year				
	1984	1980	1975	1965	1955
	- Metric Tons Total N (X 1000) -				
United States	10,111	10,348	7,803	4,208	1,779
Region					
New England States[1]	34	38	37	37	24
Middle Atlantic States[2]	259	295	243	170	111
South Atlantic States[3]	842	856	774	528	360
East North Central States[4]	2217	2209	1539	863	242
West North Central States[5]	3240	3658	2527	976	254
East South Central States[6]	602	541	510	349	246
West South Central States[7]	1354	1226	956	556	175
Mountain States[8]	636	551	415	188	69
Pacific States[9]	898	936	758	487	240

1. NE = Connecticut, Maine, Massachusetts, New Hampshire, Rhode Island, and Vermont.
2. MA = Delaware, Maryland, New Jersey, New York, Pennsylvania, and West Virginia.
3. SA = Florida, Georgia, North Carolina, South Carolina, and Virginia.
4. ENC = Illinois, Indiana, Michigan, Ohio, and Wisconsin.
5. WNC = Iowa, Kansas, Minnesota, Missouri, Nebraska, North Dakota, and South Dakota.
6. ESC = Alabama, Kentucky, Mississippi, and Tennessee.
7. WSC = Arkansas, Louisiana, Oklahoma, and Texas.
8. Mtn = Arizona, Colorado, Idaho, Montana, Nevada, New Mexico, Utah, and Wyoming.
9. Pac = California, Oregon, and Washington.

Table 8. Corn and wheat plantings, harvest tonnage, and yield in the United States from 1960 to 1984 (USDA, 1985).

Year	Plantings[1]	Harvested tonnage	Yield[2]
	hectares (x 1000)	metric tons (x 1000)	Mg/ha
		CORN	
1960	32,977	120,585	3.44
1965	26,394	126,632	4.65
1970	27,074	128,146	4.55
1975	31,881	180,270	5.42
1980	34,037	204,920	5.71
1984	32,576	236,302	6.69
		WHEAT	
1960	22,237	44,798	1.76
1965	23,231	43,505	1.78
1970	19,739	44,694	2.09
1975	30,334	70,335	2.06
1980	32,719	78,735	2.25
1984	32,081	85,829	2.61

1. Corn for grain plus silage.
2. Yield per harvested hectare.

Although total N usage has plateaued in the last five years, there have been some significant shifts in the sources of N used in the last 10 years (Tables 9 and 10). (The 1975 and 1984 years were used because CT was practiced on a very small acreage base in 1975 while in 1984 CT was used to a much larger degree throughout the U.S.). Even though AA usage has been slipping slightly, it still remains as the most common N source with over 35% of market share. Sizable increases in the use of UAN solutions and urea have occurred. Currently, UAN and urea comprise over 21 and 11% of the U.S. market, respectively. The relatively high popularity of UAN in recent years is probably related to the fact that many herbicides can be tank-mixed with UAN, thus saving one or more extra trips across the field. Ammonium nitrate usage is down 40% over this period and now comprises less than 7% of the market.

Table 9. Amounts of major N sources used in 1975 and 1984 (Hargett & Berry, 1985).

Area	An. Ammonia		UAN Soln's[1]		Urea		Am. Nitrate	
	1984	1975	1984	1975	1984	1975	1984	1975
	metric tons of N (X100)							
U.S.	35454	29890	21877	11185	11074	4699	6947	8625
Region[2]								
NE	2	2	6	9	54	37	7	28
MA	184	183	658	385	319	180	113	210
SA	305	509	3392	2684	142	53	733	1020
ENC	9565	7449	5597	2655	2234	836	367	509
WNC	16552	13797	6130	2887	3147	1114	1236	2483
ESC	7146	8373	829	591	1213	349	1663	1766
WSC	3876	3304	1813	654	2302	1109	1422	1507
Mtn.	2187	1631	1037	413	829	271	940	684
Pac.	2067	2165	2387	905	807	719	464	418

1. Assumed an average N content of 30%.
2. From Table 7.

Table 10. Percent of major N sources used in 1975 and 1984 (Hargett & Berry, 1985).

Area	An. Ammonia		UAN Soln's[1]		Urea		Am. Nitrate	
	1984	1975	1984	1975	1984	1975	1984	1975
	% of total N applied							
U.S.	35.1	38.3	21.6	14.3	11.0	6.0	6.9	11.0
Region[2]								
NE	0.4	0.6	1.6	2.5	15.6	9.9	2.1	7.5
MA	7.1	7.5	25.3	15.8	12.3	6.6	4.4	8.6
SA	3.6	6.6	40.3	34.7	1.7	0.7	8.7	13.2
ENC	43.1	48.4	25.2	17.2	10.1	5.4	1.6	3.3
WNC	51.1	54.6	18.9	11.4	9.7	4.4	3.8	9.8
ESC	11.9	16.4	13.8	11.6	20.1	6.8	27.6	34.6
WSC	28.6	34.5	13.4	6.8	17.0	11.6	10.5	15.8
Mtn.	34.4	39.2	16.3	9.9	13.0	6.5	14.8	16.4
Pac.	23.0	28.6	26.6	11.9	9.0	9.5	5.2	5.5

1. Assumed an average N content of 30%.
2. From Table 7.

These shifts in the N source market over the 1975 to 1984 period are generally observed throughout all of the U.S. regions (Tables 9 & 10). However, particular sources of N have been historically strong in certain areas and continue to remain strong. For example, AA has been the dominant source of N throughout the North-Central and Mountain states while UAN solutions have been much stronger in the South and Middle Atlantic states. Usage of UAN has increased more rapidly in the North Central, West South-Central, Mountain, and Pacific states. Urea usage has increased rather consistently in all regions except the South Atlantic and Pacific states. Declines in AA and AN use have occurred in all of the regions.

When comparing 1984 to 1975, corn acreage in the U.S. increased about 6% while a 35% increase in N rate was noted (Table 11). How much of this increase in fertilizer use is due to higher application rates associated with perceived needs of more N needed with CT tillage is not known. Perhaps the primary reason is improved management inputs (higher yielding hybrids, better pest control measures, irrigation, etc.) that lead to higher yield potential, thus requiring more N.

Table 11. Corn acreage (grain) and average N rate applied in 1975 and 1984 (Hargett & Berry, 1985).

Area	Corn planted		Avg. N rate		1975-1984 Increase in N rate
	1984	1975	1984	1975	
	Million hectares		-- kg N/ha --		%
U.S.	29.0	27.2	150	111	35
Region[1]					
NE	--	--	--	--	--
MA	1.2	1.0	NA[2]	112	--
SA	1.5	2.0	NA	139	--
ENC	10.8	9.8	158	113	40
WNC	13.2	12.5	141	105	34
ESC	1.1	1.1	NA	115	--
WSC	0.7	0.5	NA	NA	--
Mtn.	0.4	0.3	NA	125	--
Pac.	0.2	0.1	NA	NA	--

1. From Table 7.
2. NA = Not available.

Over 82% of the corn grown for grain in the U.S. is planted in the North-Central states (Table 11). This same region uses over 54% of the total N in the U.S. and 74, 54, 48, and 23% of the AA, UAN, urea, and AN, respectively.

Wheat acreage decreased 4% from 1975 to 1984, but a 62% increase in N rate occurred (Table 12). Much of that increase was due to higher yielding varieties with greater stem strength. Consequently, N rates could be raised to increase yield without causing lodging. Another factor could be the shift away from the alternate-year fallow rotation used in some of the Mountain states. As continuous small grains were planted, fertilizer N requirements became higher to maximize both yield and economic return.

Table 12. Wheat acreage and average N rate applied in 1975 and 1984 (Hargett & Berry, 1985).

Area	Wheat planted		Avg. N rate		1975–1984 Increase in N rate
	1984	1975	1984	1975	
	Million hectares		-- kg N/ha --		%
U.S.	27.1	28.2	53	32	62
Region[1]					
NE	--	--	--[2]	--	--
MA	0.2	0.3	NA[2]	NA	--
SA	0.9	0.4	NA	NA	--
ENC	1.9	2.5	74	54	38
WNC	12.3	13.3	44	28	56
ESC	0.4	0.8	NA	NA	--
WSC	4.9	5.2	67	31	114
Mtn.	4.2	4.0	36	16	128
Pac.	1.8	2.1	85	73	17

1. From Table 7.
2. NA = Not available.

Approximately 79% of the wheat grown in the U.S. is planted in West North-Central, West South-Central, and Mountain states (Table 12). These regions use over 52% of the N in the U.S. and 64, 41, 57 and 52% of the AA, UAN, urea, and AN, respectively.

SUMMARY

Conservation tillage leaves significant amounts of plant residues on the soil surface which can greatly affect water intake and N loss mechanisms, e.g., volatilization, leaching, and denitrification. Volatilization losses of N from surface-applied urea-containing N fertilizers can be substantial. Moreover, for a variety of reasons the trend amoung producers has been to move away from anhydrous ammonia toward urea and UAN, which are often surface-applied. This paradox requires the development of better injection/incorporation techniques that allow growers to manage their N better in the future. Presently, agronomists throughout the U.S. recognize this volatilization problem and many make recommendations to guard against these losses. In some cases, higher rates of N are recommended with CT compared to conventional tillage. This could lead to increased amounts of NO_3 leached toward the groundwater if the N is not taken up by the plant, volatilized, immobilized or denitrified.

Under drier, non-irrigated conditions of the Great Plains, CT is generally thought to reduce NO_3 losses to the groundwater because of greater water use efficiency resulting in greater N uptake and higher yield. In the more humid areas of the U.S. a consensus among agronomists was not reached as to the relationship of CT to NO_3 in the groundwater. The role of CT on the fate of N in a cropping system needs to be more clearly defined with further research; especially with respect to immobilization, mineralization, and water flow characteristics.

REFERENCES

Baker, J. L., and J. M. Laflen. 1983. Water quality consequences of conservation tillage. J. Soil and Water Cons. 38:186-193.

Baker, J. L., T. S. Colvin, S. J. Marley, M. Dawelbiet. 1985. Improved fertilizer management with a point-injector applicator. Paper 85-1516. Am. Soc. Agr. Eng., St. Joseph, Mich.

Bandel, V. A., S. Dzienia, and G. Stanford. 1980. Comparison of N fertilizers for no-till corn. Agron. J. 72:337-341.

Bandel, V. A. 1984. Maximizing N efficiency in no-till corn. Solutions 28:36-42.

Bandel, V. A. 1986. Fertilization of no-till corn - Part II. The agronomist (Univ. of Maryland) Vol. 23 No. 5:9-10.

Boswell, F. C., J. J. Meisinger, and N. L. Case. 1985. Production, marketing and use of nitrogen fertilizers. p. 229-292. In O. P. Engelstad (ed.) Fertilizer technology and use. American Society of Agronomy, Crop Science Society of America and Soil Science Society of America, Madison, WI.

Culley, J. L. B. 1986. Water regimes and shear strengths of a Typic Haplaquoll under conventional and no tillage. Ph.D. diss. Univ. of Minnesota, St. Paul.

Dawelbeit, M., J. L. Baker, and S. J. Marley. 1981. Design and development of a point-injector for liquid fertilizer. Paper 81-1010. Am. Soc. Agr. Eng., St. Joseph, Mich.

Dickey, E. C., P. J. Jasa, A. J. Jones, and D. P. Shelton. 1986. Conservation tillage systems for row crop production. p. 1-7. In E. C. Dickey (ed.) Conservation tillage for row crop production. Conservation Tillage Proc. No. 5. Univ. of Nebraska, Lincoln, NE.

Fox, R. H., and L. D. Hoffman. 1981. The effect of N fertilizer source on grain yield, N uptake, soil pH, and lime requirement in no-till corn. Agron. J. 73:891-895.

Frye, W. W., R. L. Blevins, L. W. Murdock, L. L. Wells, and J. H. Ellis. 1981. Effectiveness of nitrapyrin with surface-applied fertilizer nitrogen in no-tillage corn. Agron, J. 73:287-289.

Griffith, D. R. 1974. Fertilization and no-plow tillage. In Proc. Indiana Plant Food and Agric. Chem. Conf., Purdue Univ., West Lafayette, Ind. 17-18 Dec. Purdue Univ., West Lafayette, IN.

Hargett, N. L., and J. T. Berry. 1985. 1984 Fertilizer summary data. TVA/OACD-85/10. Bul. Y-189. Nat. Fert. Devel. Ctr. Tennessee Valley Authority, Muscle Shoals, AL 35660.

Horton, R. 1986. Does anion leaching occur in Iowa soils? In Proc. Iowa Fert. and Agric. Chem. Dealers Conf., Iowa State Univ., Ames, Iowa. 14-15 Jan. Des Moines, IA.

Lindstrom, M. J., W. B. Voorhees, and G. W. Randall. 1981. Long-term tillage effects on interrow runoff and infiltration. Soil Sci. Soc. Am. J. 45:945-948.

Mengel, D. B., D. W. Nelson, and D. M. Huber. 1982. Placement of nitrogen fertilizers for no-till and conventional till corn. Agron. J. 74:515-518.

Randall, G. W. 1984. Efficiency of fertilizer nitrogen use as related to application methods. p. 521-533. In R. D. Hauck (ed.) Nitrogen in crop production. American Society of Agronomy, Crop Science Society of America, and Soil Science Society of America, Madison, WI.

Randall, G. W., J. W. Bauder, W. E. Lueschen, and J. B. Swan. 1980. Continuous corn tillage study. p. 134-141. In A report on field research in soils. Univ. of Minnesota Agric. Exp. Stn. Misc. Pub. 2.

Randall, G. W., and P. L. Kelly. 1986. Nitrogen loss to tile lines as affected by tillage. p. 125-127. In A report on field research in soils. Univ. of Minnesota Agric. Exp. Stn. Misc. Pub. 2 (revised).

Randall, G. W., and D. K. Langer. 1982. Nitrogen efficiency as affected by ridge-planting. p. 136-139. In A report on field research in soils. Univ. of Minnesota Agric. Exp. Stn. Misc. Pub. 2(revised).

Randall, G. W., K. L. Wells, and J. J. Hanway. 1985. Modern techniques in fertilizer application. p. 521-560. In O.P. Engelstad (ed.) Fertilizer technology and use. American Society of Agronomy, Crop Science Society of America, and Soil Science Society of America, Madison, WI.

Stevenson, F.J. 1982. Origin and distribution of nitrogen in soil. In F. J. Stevenson (ed.) Nitrogen in agricultural soils. Agronomy 22:1-42. Am. Soc. of Agron., Madison, WI.

Thomas, G. W., R. L. Blevins, R. E. Phillips, and M. McMahon. 1973. Effect of killed sod mulch on nitrate movement and corn yield. Agron. J. 65:736-739.

Touchton, J. T., and W. L. Hargrove. 1982. Nitrogen sources and methods of application for no-tillage corn production. Agron. J. 74:823-826.

USDA Staff. 1985. Agricultural statistics. U. S. Government Printing Office, Washington, DC.

Wells, K. L. 1984. Nitrogen management in the no-till system. p. 535-550. In R. D. Hauck (ed.) Nitrogen in crop production. American Society of Agronomy, Crop Science Society of America, and Soil Science Society of America, Madison, WI.

CHAPTER 4

OVERVIEW OF RURAL NONPOINT POLLUTION IN THE LAKE ERIE BASIN

D. B. Baker,
Heidelberg College, Tiffin, Ohio

INTRODUCTION

Water pollution associated with land use activities (i.e., nonpoint source pollution) has received particularly detailed study within the Great Lakes Basin. Through a series of U.S.-Canadian investigations coordinated by the International Joint Commission's Pollution from Land Use Activities Reference Group (PLUARG), a comprehensive overview of nonpoint source pollution in the Great Lakes was developed (International Joint Commission, 1978, 1980, 1983). These studies revealed that land use activities adversely impact great lakes water quality. Agricultural land use was singled out as a major source of sediments, nutrients and pesticides impacting several regions, including Green Bay, Saginaw Bay and much of the western and central basins of Lake Erie. These studies indicated that, although the land area draining into Lake Erie occupies only 11.5% of the total land area in the Great Lakes Basin, Lake Erie tributaries carried 58% of the total tributary suspended solids load entering the Great Lakes (International Joint Commission, 1978). Maps of unit area phosphorus yields for the Great Lakes indicated that the largest aggregation of lands with high unit area phosphorus yields occurred in the watersheds draining into the western and central basins of Lake Erie. These high sediment and phosphorus losses are associated with the intensive row crop agriculture which dominates land use in large portions of the Lake Erie Basin. Consequently, agricultural nonpoint

Effects of Conservation Tillage on Groundwater Quality: Nitrates and Pesticides, Terry J. Logan et al., eds. © 1987 Lewis Publishers, Inc., Chelsea, Michigan 48118. Printed in USA.

pollution has been studied most extensively in the Lake Erie Basin.

Much of the detailed study in the Lake Erie Basin was conducted as part of the U.S. Army Corps of Engineers' Lake Erie Wastewater Management Study (LEWMS) (U.S. Army Corps of Engineers, 1982). This study included the development of a detailed geographical information system for the entire United States portion of the Lake Erie Basin (Adams et al., 1982) as well as detailed water quality studies (Baker, 1984, 1985). The LEWMS program was coordinated with both the PLUARG studies and the Areawide Waste Treatment Management planning studies conducted under Section 208 of the Federal Water Pollution Control Act Amendments of 1972 (Public Law 92-500).

As it became evident in the above studies that agriculture was a major source of phosphorus entering Lake Erie, ways to reduce agricultural phosphorus loading were examined. Conservation tillage was quickly identified as a potentially effective means of reducing erosion and the associated suspended sediment and particulate phosphorus loading into Lake Erie. The agronomic suitability of conservation tillage for Lake Erie Basin soils was then evaluated in a series of demonstration studies. The first of these demonstrations was located in the Honey Creek Watershed of the Sandusky River Basin as part of the LEWMS study. The success of the Honey Creek Demonstration Project (Honey Creek Joint Board of Supervisors, 1982) led to U.S. EPA-supported conservation tillage demonstration programs in 31 counties of the Lake Erie Basin (Morrison, 1984). The major objectives of these demonstration studies were to acquaint as many farmers as possible with conservation tillage techniques, to develop local data comparing conventional tillage and conservation tillage in terms of crop yields and production costs, and to accelerate area-wide adoption of conservation tillage. These demonstration studies have confirmed that, for many Lake Erie Basin soils, conservation tillage can provide either equivalent or increased profits in comparison with conventional tillage (Anonymous, 1985).

In 1983, through a Supplement to Annex 3 of the Great Lakes Water Quality Agreement of 1978, the U.S. and Canada agreed to reduce phosphorus loading to Lake Erie by an additional 2,000 metric tons per year beyond the reductions achievable by reducing major municipal point source phosphorus loading to 1 mg/L P in the effluents. The U.S. phosphorus reduction strategy (Great Lakes Phosphorus Task Force, 1985), as well as those of the individual states, (e.g., see Ohio EPA, 1985) is focusing on conservation tillage as a major tool to reduce phosphorus loading to the lake. Implementation of agricultural phosphorus load reduction programs should consequently consist of continuing and expanding programs to aid farmers in adopting conservation tillage.

While the erosion reduction benefits of conservation tillage are well documented, at least at the level of plot and field sized studies, much concern exists regarding the possibility that conservation tillage could aggravate other water quality problems, especially the contamination of surface and ground-water by nitrates and pesticides (Crosson, 1981; Hinkle, 1983). This workshop reflects that concern on the part of the U.S. EPA's Great Lakes National Program Office. Since our laboratory has been involved in comprehensive nonpoint source pollution studies in the Lake Erie Basin for many years, I have been asked by the organizers of this workshop to present an overview of nitrate and pesticide contamination in this region's surface water, as well as data on sediment and phosphorus loads in the rivers of this region.

Lake Erie Tributary Monitoring Programs

The Water Quality Laboratory at Heidelberg College has been responsible for implementing the major tributary monitoring programs associated with the PLUARG, LEWMS, "208" and conservation tillage demonstration projects in the Lake Erie Basin. Consistent techniques, involving the use of automatic samplers for storm event sampling of tributaries and automated analytical procedures for chemical analyses, have been used throughout these studies and have been described in detail by Baker (1984) and Kramer and Baker (1985). All of the sampling is conducted at U.S. Geological Survey stream gaging stations and the U.S. Geological Survey provides both hourly gage height data and stage-discharge tables for each station. The scope of the monitoring programs is summarized in Table 1.

The analytical program for each sample noted in Table 1 includes analyses for suspended solids, total phosphorus, soluble reactive phosphorus, nitrate plus nitrite nitrogen, chloride, conductivity and, since 1981, Total Kjeldahl nitrogen. Also since 1981, forms of bioavailable phosphorus and several currently-used pesticides have been analyzed in selected samples. All analyses are performed on discrete samples rather than on composited samples. Analytical results are available in the U.S. EPA's STORET system.

The watersheds upstream from each sampling station range in area from 8.8 km^2 to 16,359 km^2. This variation in watershed area supports analysis of "scale effects" of watershed size (and associated stream order) on the concentration and loading patterns of agricultural chemicals as they move through stream systems. The land use and soil characteristics for each watershed are available from the LEWMS geographical data base (Resource Management Associates, 1979). Land use is summarized in Table 2. Most of the soils are fine textured and derived from glacial till or lake bed materials. Tile drainage systems are used extensively within this region.

Table 1. Wastershed Areas, Mean Annual Discharges, Period of Chemical Sampling and Number of Samples Analyzed for the Lake Erie Tributary Monitoring Program

Transport Stations	U.S. Geological survey station number	Drainage Area km^2	Mean Annual Discharge Years of Record	m^3/s	cm	Chemical Sampling Period	Number of Samples Analyzed
			Sandusky River Stations				
Fremont	04198000	3,240	57	27.75	27.0	1974–84	4590+
Mexico	04197000	2,005	55	16.62	26.2	1974–81	2178
Upper Sandusky	04196500	722	57	6,967	28.5	1974–81	2973
Bucyrus	04196000	230	40	2.461	33.8	1974–81	2998
			Sandusky River Tributaries				
Wolf Cr., East	04192450	213	5	1.82	27.0	1976–81	2425
Wolf Cr., West	04197300	171.5	5	1.34	24.6	1976–81	2419
Honey Cr., Melmore	04197100	386	7	3.908	32.0	1976–84	4595+
Honey Cr., New Wash.	04197020	44.0	3.908	(0.445)[1]	(32.0)[1]	1979–81; 83–84	2271+
Tymochtee Cr.	04196800	593	19	4.956	26.3	1974–81	2471
Broken Sword Cr.	04196200	217	5	2.45	35.5	1976–81	2512
			Other Lake Erie Tributaries				
Maumee River	04193500	16,359	58	139.5	26.8	1975–80;82–84	3154
Raisin	04176500	2,699	43	19.85	23.2	1982–84	805+
Cuyahoga	04208000	1,831	52	23.14	39.8	1981–84	1380+
Portage	04195500	1,109	51	9.091	25.9	1974–78	1856
Huron	04199000	961	31	8.496	27.9	1974–79	2027

[1] Extrapolated from Honey Creek at Melmore

Table 2. Watershed Land Use for the Lake Erie Tributary Nutrient and Sediment Transport Stations

Transport Stations	Cropland	Pasture	Forest	Water	Other
	%				
Sandusky River Stations					
Fremont	79.9	2.3	8.9	2.0	6.8
Mexico	80.3	2.3	8.7	2.1	6.6
Upper Sandusky	78.0	3.4	9.1	2.0	7.5
Bucyrus	73.3	4.9	9.4	2.1	10.2
Sandusky River Tributaries					
Wolf Cr., East	81.9	2.7	6.3	2.0	7.0
Wolf Cr., West	83.3	1.4	4.7	3.1	7.6
Honey Cr., Melmore	82.6	0.6	10.0	0.5	6.3
Honey Cr., New Wash.	89.1	–	7.5	–	3.4
Tymochtee Cr.	84.0	1.2	7.6	2.3	4.8
Broken Sword Cr.	84.7	1.4	8.5	1.3	4.1
Other Lake Erie Tributaries					
Maumee River	75.6	3.2	8.4	3.5	9.4
Raisin	67.1	6.8	9.0	3.0	14.1
Cuyahoga	4.2	43.1	29.1	3.0	20.6
Portage	85.5	3.6	5.6	0.9	4.3
Huron	75.3	3.5	12.5	2.2	6.4

Material Transport During Storm Events

Quantitative studies of material export from agricultural watersheds essentially involve detailed studies of individual runoff events or floods. In the rivers of this region, a recurring pattern of concentration changes usually occurs during runoff events. This pattern is illustrated in Figure 1. Peak concentrations of suspended sediment and total phosphorus preceed the peak discharge while peak nitrate concentrations trail the peak discharge. Peak herbicide concentrations occur between the times of peak sediment and peak nitrate concentrations. The trailing peaks for nitrate probably reflect the delayed arrival, at the monitoring station, of tile drainage water relative to surface runoff water. The bulk of the nitrate export from fields of this region is through the tile systems. The advanced peaks for sediment and total phosphorus concentrations probably reflect some combination of material routing from individual fields to and through stream systems and the deposition and resuspension of sediment with the passage of the flood wave through the stream system (Baker, 1984; Verhoff et al., 1978). The pattern for herbicides may simply reflect that surface runoff water transports herbicides from fields into streams.

While the sequence of peak concentrations and flow illustrated in Figure 1 is common, the actual concentrations and flows vary tremendously from storm to storm at a given station as well as from station to station. Examples of storm to storm variability have been presented by Baker (1984).

Annual Chemographs

The sampling programs are conducted throughout the year at each station and are summarized by water year (October 1 through September 30). Examples of annual chemographs for one of the transport stations (the Maumee River) are shown in Figure 2. The type of data illustrated in Figure 2 is available for the Sandusky River at Fremont beginning with the 1975 Water Year (October 1, 1974-September 30, 1975) and for other stations for the years covered by the chemical sampling period (Table 1). Depending on the watershed size and weather, from 400 to 650 samples are analyzed each year to characterize stream transport of sediments and nutrients.

Annual and Seasonal Variability in Loadings

Data of the type illustrated in Figure 2 can serve as a basis for studying either pollutant loading and transport in the rivers or for studying pollutant concentration patterns and durations. Given multiple years of study, the extent of annual variability as well as the seasonal characteristics of material loadings and concentrations can also be examined. Figure 3

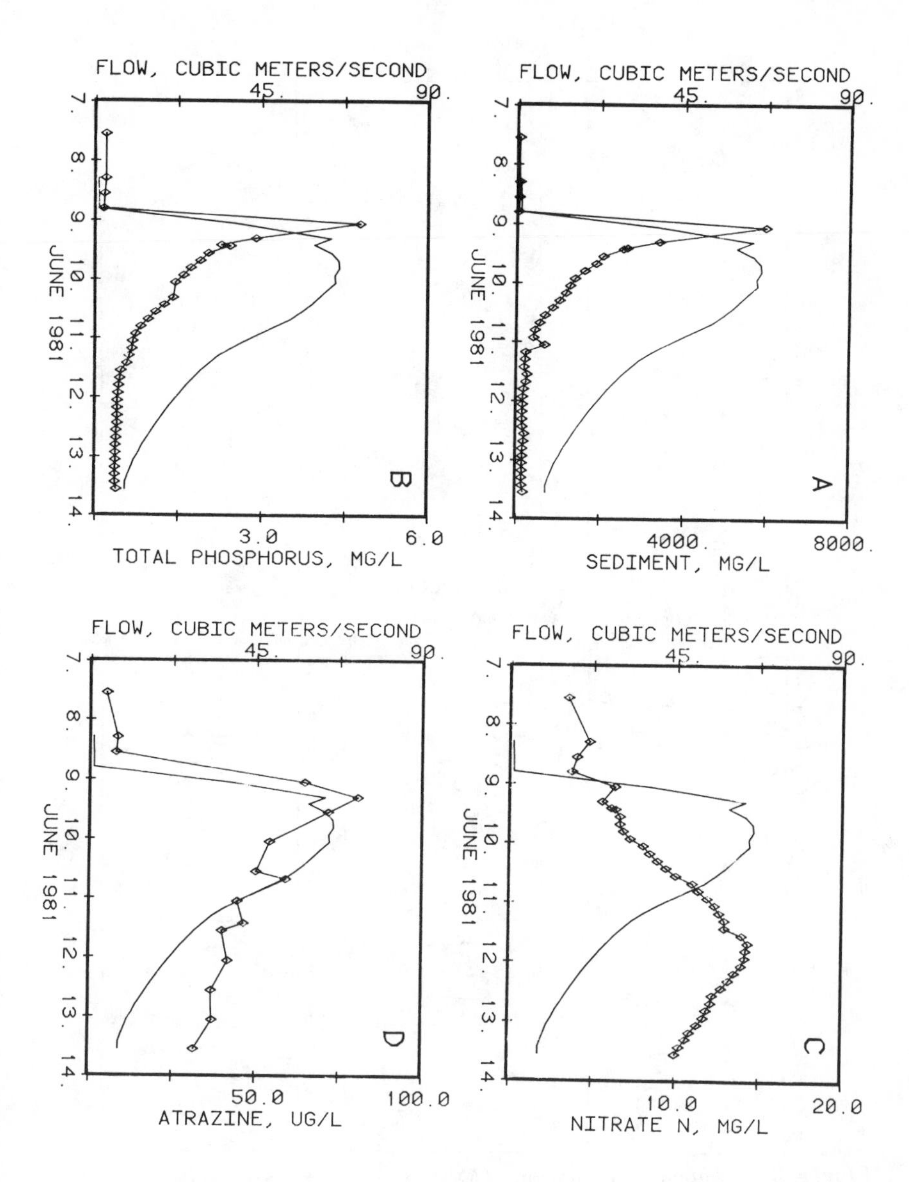

Figure 1. Concentration patterns in relationship to storm hydrographs for sediments (A), total phosphorus (B), nitrate + nitrite-nitrogen (C), and atrazine (D) in a typical runoff event for the Honey Creek Watershed at Melmore, Ohio in northwestern Ohio.

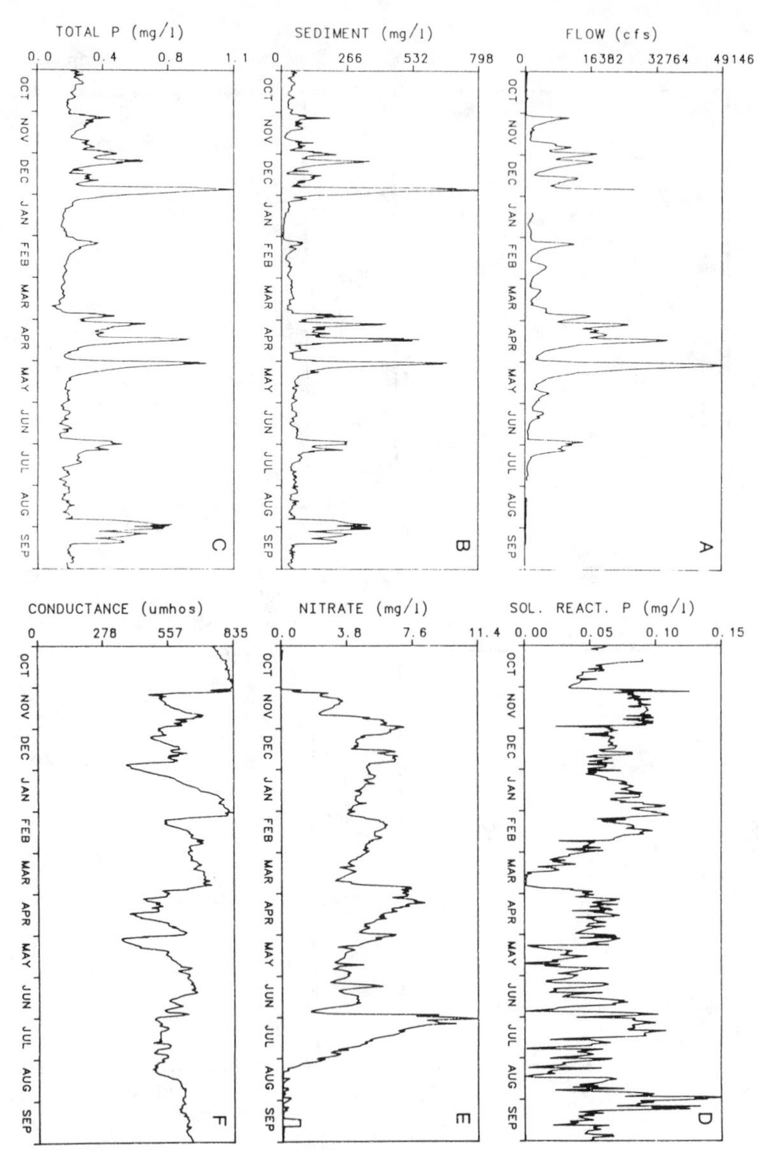

Figure 2. Annual hydrograph (A) and concentration patterns of suspended sediment (B), total phosphorus (C), soluble reactive phosphorus (D) nitrate + nitrite-nitrogen (E), and conductivity (F) for the Maumee River at Bowling Green, Ohio water intake for the 1983 water year.

illustrates both the annual variability and seasonal composition of the export of agricultural chemicals from the Sandusky River. The variability in annual loads apparent in Figure 3 largely represents the effects of variable weather conditions from year to year. During the 1975-1984 period there were no significant changes in either land use or agricultural management practices that could account for the fluctuations in annual loads of sediments and nutrients evident in Figure 3. During this time period, the use of no-till and reduced till in the Sandusky Basin gradually increased from essentially zero up to 15% to 18% of the corn and soybean acreage, respectively. Total rainfall also does not account for much of the variability. To account for the variability, rainfall amounts and intensities associated with individual runoff events will have to be tracked. It is evident from Figure 3 that short term studies of one to three years duration could give rather misleading estimates of mean annual loads. It is also evident that efforts to document the effectiveness of changing management practices in reducing nutrient and sediment loading will require compensation for the variability associated with weather conditions.

In Table 3 the percentages of the annual export occurring in fall (October - December), winter (January - March), spring (April - June), and summer (July - September) are shown for three watersheds of differing sizes. Flux weighted mean concentrations of nutrients and sediments for the corresponding seasons and stations are shown in Table 4. For all three watersheds, the seasonal distribution of discharges is similar, with the January-March period accounting for the largest proportion of discharge even though this period receives the least rainfall. The least amount of discharge occurs in the July-September period. For the smaller Honey Creek watershed, the spring period accounts for the bulk of the sediment export while the winter period is less important. The reverse situation occurs for the larger Maumee watershed. These seasonal differences in suspended solids export are also reflected in the seasonal flux weighted mean concentrations of suspended solids. The winter and spring periods have similar importance in terms of nitrate-nitrogen export for all of the watersheds. However, the flux weighted concentrations of nitrate are much higher in the spring than in the winter for all three watersheds.

Of particular importance from a water quality standpoint are the high proportions of the soluble reactive phosphorus loading that occur in winter. Since soluble reactive phosphorus has a high bioavailability, the winter loading of this phosphorus fraction could be particularly important as a phosphorus source for Lake Erie. More information is needed on the sources and possible control of these winter-time soluble phosphorus loads.

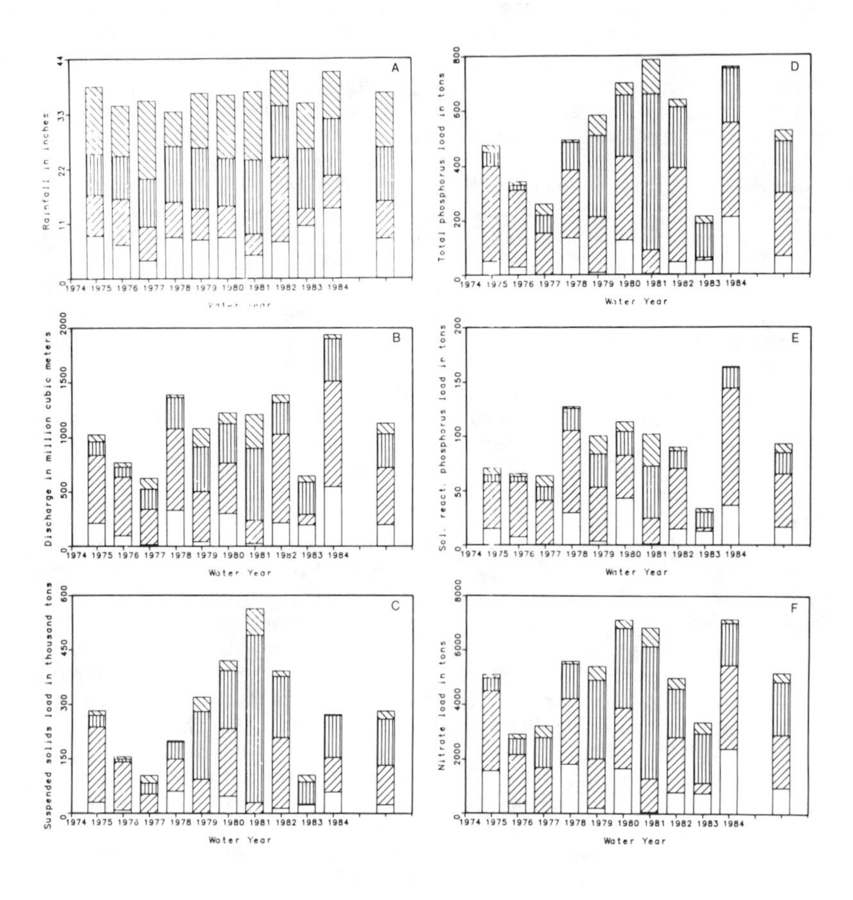

Figure 3. Annual variability and seasonal patterns of rainfall (A), discharge (B) and export of suspended solids (C), total phosphorus (D), soluble reactive phosphorus (E) and nitrate + nitrite-nitrogen (F) for the Sandusky River at Fremont, Ohio during the 1975–1984 water years. Each bar reflects the seasonal loads beginning with the October-December loads at the base (clear), followed by the January-March loads, the April-June loads (vertical hatching) and the July-September loads.

Table 3. Seasonal Distribution of Rainfall, Discharge and Nutrient and Sediment Export from Three Northwest Ohio Watersheds of Varying Sizes

	Percent of Mean Annual Load			
	Oct-Dec	Jan-March	April-June	July-Sept
Rainfall				
Honey Creek [see Sandusky R.]				
Sandusky R.	21.4	20.2	29.3	29.0
Maumee R.	22.3	17.9	31.2	28.6
Discharge				
Honey Creek	20.2	41.6	30.5	7.7
Sandusky R.	17.6	46.4	27.3	8.7
Maumee R.	17.6	40.7	33.8	8.0
Suspended Sediment				
Honey Creek	7.1	27.2	57.2	8.4
Sandusky R.	8.9	38.4	44.8	7.9
Maumee R.	16.9	42.4	37.3	3.4
Total Phosphorus				
Honey Creek	14.2	34.5	43.8	7.5
Sandusky R.	12.6	44.4	35.5	7.5
Maumee R.	20.0	43.7	32.5	3.8
Soluble Reactive Phosphorus				
Honey Creek	23.9	41.3	24.6	10.1
Sandusky R.	17.9	52.5	20.7	8.9
Maumee R.	21.4	48.8	21.2	5.5
Nitrate + Nitrite-Nitrogen				
Honey Creek	20.3	33.2	39.0	7.5
Sandusky R.	18.8	37.9	37.0	6.3
Maumee R.	24.3	34.9	36.1	4.7

Table 4. Seasonal and Annual Flux Weighted Mean Concentration of Sediments and Nutrients for the Period of Record

	Flux Weighted Mean Concentration (mg/L)				
	Oct–Dec	Jan–March	April–June	July–Sept	Overall
Suspended Solids					
Honey Creek	72	133	381	221	203
Sandusky R.	125	206	409	226	249
Maumee R.	179	205	272	140	216
Total Phosphorus					
Honey Creek	0.294	0.346	0.598	0.407	0.417
Sandusky R.	0.332	0.444	0.603	0.402	0.464
Maumee R.	0.445	0.473	0.531	0.360	0.479
Soluble Reactive Phosphorus					
Honey Creek	0.088	0.074	0.060	0.098	0.074
Sandusky R.	0.083	0.093	0.062	0.085	0.082
Maumee R.	0.092	0.095	0.071	0.092	0.087
Nitrate + Nitrite–Nitrogen					
Honey Creek	4.84	3.85	6.16	4.67	4.82
Sandusky R.	4.87	3.73	6.19	3.35	4.57
Maumee R.	5.25	3.76	5.87	4.39	4.82

Mean Annual Loads, Unit Area Loads, and Sediment Delivery Ratios

In Table 5 the mean annual loads and relative standard deviations are shown for sediments and nutrients for three of the transport stations. The mean loads are simply the averages for the monitored years and are not corrected for long term discharge conditions. Sediment export from the smallest watershed (Honey Creek) was the most variable. Soluble constituents had less variability from year to year than did suspended solids or total phosphorus for Honey Creek and the Sandusky River but not for the Maumee River.

The unit area yields of sediments and nutrients shown in Table 6 were derived by dividing the mean annual load (Table 5) by the entire drainage area upstream from the transport station. Rast and Lee (1985) estimated that the average unit area yields of total phosphorus and total nitrogen from agricultural lands in the U.S. were, respectively, 0.49 and 5.04 kg/ha/yr. The unit area yields for total phosphorus and nitrate+nitrite-nitrogen for northwestern Ohio watersheds are three times higher than these averages.

In the case of phosphorus, point source inputs within the watersheds also contribute to the river systems. For the Honey Creek, Sandusky and Maumee watersheds, these point source phosphorus inputs, expressed on a unit area basis, are only 0.09, 0.14 and 0.20 kg/ha/yr and thus can account for only 6.6%, 8.6% and 12.8% of the total phosphorus export, respectively. These calculations assume 100% delivery of point source-derived phosphorus through the stream systems and consequently may overestimate the contribution of point sources to the phosphorus yields (Baker, 1980). Overall, rural nonpoint sources account for approximately 60% of the total phosphorus loading into Lake Erie.

As part of the LEWMS study, gross erosion rates were calculated for many Lake Erie watersheds (Logan et al., 1982). Average gross erosion rates for three of the study watersheds are listed in Table 6. These gross erosion rates, coupled with the sediment export data, indicate that the sediment delivery ratios for these watersheds are approximately 10%.

Lake Erie Basin and Chesapeake Bay Basin Comparisons

The large magnitude of agricultural pollution in the Lake Erie Basin is evident when compared to data from the Chesapeake Bay Region (Macknis, 1985; Smullen et al., 1982). While the populations of both areas are the same, the drainage area of Chesapeake Bay is approximately three times larger than that of Lake Erie (Table 7). River loadings of sediment, total phosphorus and total nitrogen are, however, much larger for Lake Erie tributaries. Consequently, the unit area loads of sediment, total phosphorus and total nitrogen are 6.4, 5.2 and 4.2 times higher, respectively, than those for Chesapeake Bay watersheds. These higher unit area loads for Lake Erie

Table 5. Means and Coefficients of Variation for Annual Discharge and for Export of Sediments and Nutrients from Three Northwestern Ohio Watersheds of Varying Sizes

Watershed (Years of data)	Discharge $10^6 m^3$	Suspended Solids 10^3 metric tons	Total Phosphorus metric tons	Soluable Reactive Phosphorus metric tons	Nitrate+ Nitrite- Nitrogen metric tons
Honey Creek (9 years)	125.0 ± 26%	25.4 ± 71%	52.1 ± 42%	9.2 ± 29%	604 ± 24%
Sandusky R. (10 years)	1,133 ± 35%	282.5 ± 52%	526 ± 38%	93.3 ± 40%	5,180 ± 31%
Maumee R. (6 years)	5,332 ± 26%	1,154 ± 25%	2,551 ± 17%	464.9 ± 33%	25,700 ± 20%

Table 6. Unit Area Yields of Sediments and Nutrients for the Period of Record, Average Gross Erosion Rates, and Average Sediment Delivery Percentages for Three Northwestern Ohio Watersheds

	Average Gross Erosion Rate mt tons/ha	Sediment Yield mt tons/ha	Average Sediment Delivery As Percent	Total Phosphorus Yield kg/ha	Soluble Reactive Phosphorus Yield kg/ha	Nitrite + Nitrite-N Yield kg/ha
Honey Creek	6.86	0.65	9.4	1.35	0.24	15.7
Sandusky River	8.25	0.87	10.5	1.62	0.29	16.0
Maumee River	6.84	0.70	10.2	1.56	0.28	15.7

Table 7. Comparison of the Lake Erie Basin and Chesapeake Bay Basin with Respect to Population, Drainage Areas and Tributrary Pollutant Loads

Parameter	Lake Erie Basin	Chesapeake Bay Basin
Population	14,000,000	14,000,000
Land Area, km^2	56,980	165,800
River Sediment Loads		
mtons/yr	6,531,000	3,005,800
kg/ha/yr	1,150	181
River Phosphorus Loads		
metric tons/yr	8,400	4,659
kg/ha/yr	1.47	0.28
River Nitrogen Loads		
metric tons/yr	111,670	77,584
kg/ha/yr	19.6	4.67

watersheds are associated with the intensive row crop agriculture which dominates much of the Lake Erie watershed. The higher population densities coupled with intensive agricultural land use put particularly heavy pressure on the water and soil resources of the Lake Erie Basin.

Nitrate Contamination of Surface Waters and Drinking Waters

In northwestern Ohio, as elsewhere in the Midwest, several municipalities withdraw water directly from rivers for public water supplies. Since conventional water treatment procedures do not remove nitrates, the nitrate concentrations present in the rivers are also present in the finished water supplies. The nitrate concentrations in area rivers frequently exceed the drinking water standard of 10 mg/L nitrate-nitrogen, usually during the May-July period. In the case of the Sandusky River, which supplies drinking water for both Fremont and Tiffin, Ohio, the nitrate standard has been exceeded every year since the onset of our monitoring program in 1975. In 1985, the standard was exceeded continuously for 30 days.

The effects of watershed size on the nitrate concentration patterns in area streams and rivers are illustrated in the concentration duration curves of Figure 4. The smaller tributary (Honey Creek) has higher peak nitrate concentrations than the Maumee, but intermediate concentrations persist for longer periods in the Maumee. For the Sandusky River, nitrates exceed the standard 4.1% of the time, but since these occurrences are always in the months of May, June or July, the standard is exceeded 16% of the time during these months. For the Sandusky, nitrates were in the range of 7-10 mg/L for about 12% of the time. If conservation tillage increases infiltration and, consequently, the proportion of stream water derived from tile effluents, it is likely that the percentage of time nitrates exceed the drinking water standard will increase.

Pesticides in Rivers and Public Water Supplies

During runoff events following pesticide applications, many currently-used pesticides are present in area rivers. Some typical storm event concentration patterns of herbicides, along with nitrates and discharge, are shown in Figure 5. Peak annual concentrations for 1983, 1984 and 1985 at three transport stations are shown in Table 8. Time weighted mean concentrations for the April 15-August 15 period for the same pesticides are shown in Table 9.

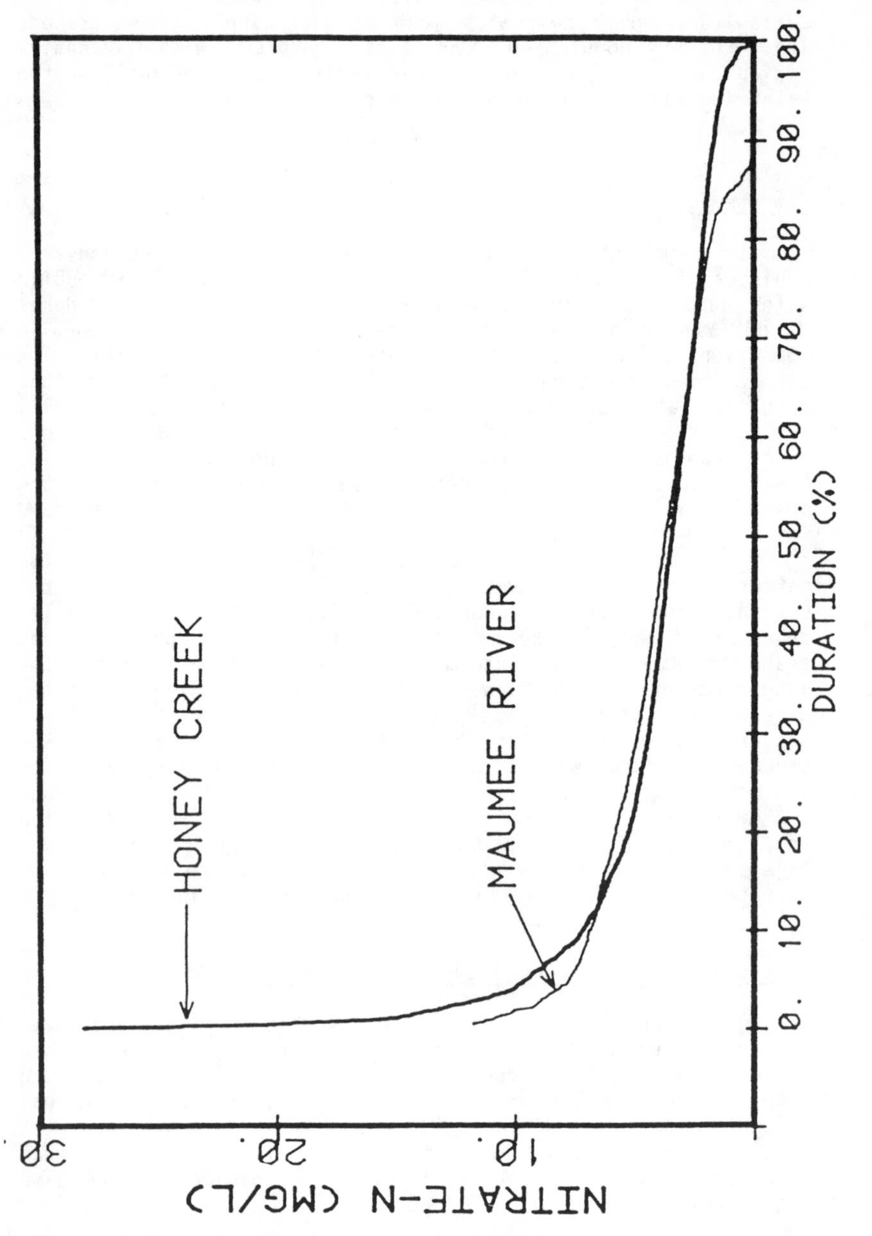

Figure 4. Comparison of the concentration-duration curves for nitrate + nitrite-nitrogen for the Maumee and Honey Creek watersheds.

Table 8. Peak Observed Concentrations of Pesticides at Three Northwestern Ohio Transport Stations

	Honey Creek			Sandusky River			Maumee River		
	1983	1984	1985	1983	1984	1985	1983	1984	1985
Simaxine	nd	1.05	0.48	0.01	1.25	0.98	nd	0.69	0.62
Carbofuran	0.43	5.12	2.40	0.50	1.41	1.24	0.48	2.42	0.59
Atrazine	17.48	32.22	20.17	7.97	8.73	19.46	5.42	11.71	6.21
Terbufos	0.016	nd	0.041	nd	nd	0.044	0.03	0.016	0.010
Fonofos	nd	nd	0.010	0.03	nd	0.049	nd	0.034	0.014
Metribuzin	3.42	3.41	4.78	2.45	4.37	3.09	4.20	5.74	1.65
Alachlor	8.87	22.89	17.32	4.92	9.10	16.97	7.49	18.38	3.61
Linuron	4.30	1.93	4.73	1.03	0.42	3.09	0.39	1.38	0.34
Metolachlor	23.42	30.81	23.45	16.70	16.92	28.38	7.03	11.95	5.71
Cyanazine	2.23	4.88	6.72	1.39	3.33	2.72	1.94	9.96	1.25

[1] nd = not detected

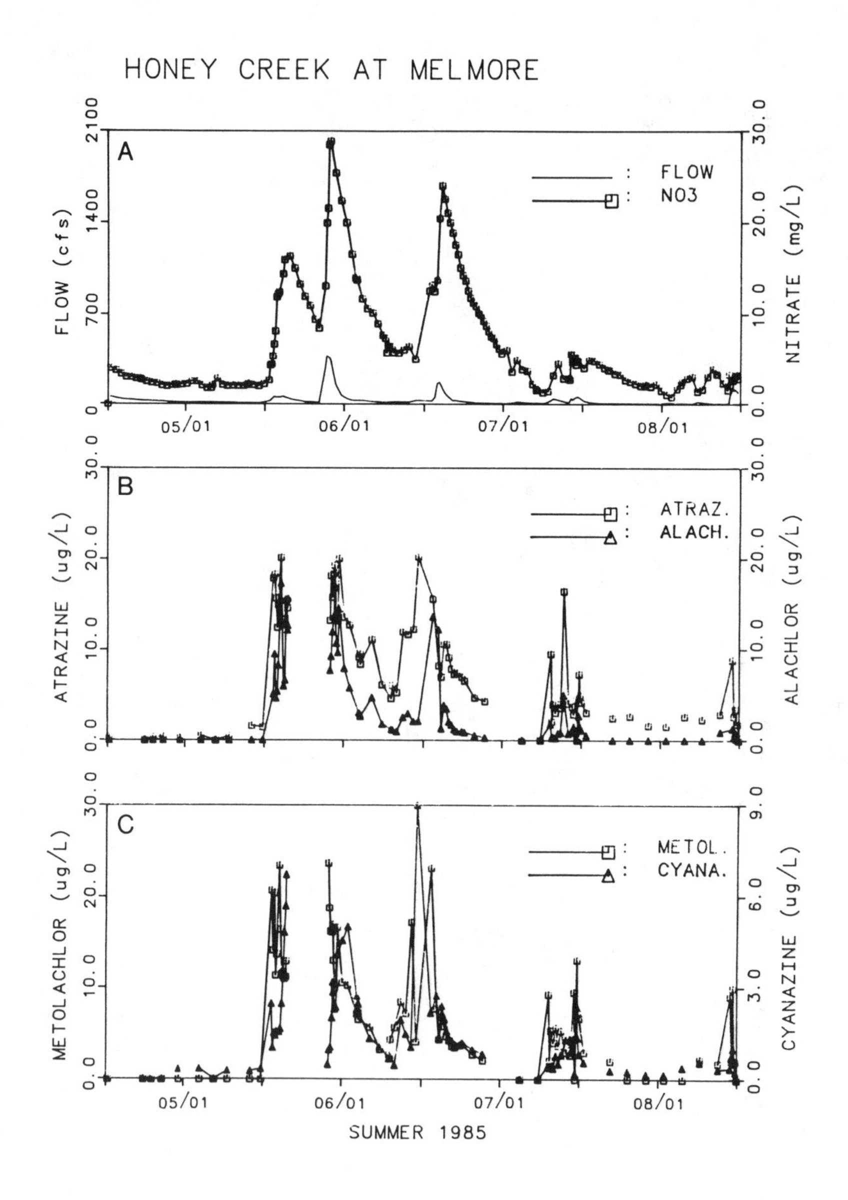

Figure 5. Spring runoff patterns for discharge and nitrate (A), atrazine and alachlor (B) and metolachlor and cyanazine (C) for the Honey Creek watershed at Melmore in 1985.

Most of the pesticides shown in Tables 8 and 9 are transported primarily in the dissolved form and are neither broken down nor removed during conventional water treatment procedures. Consequently, for cities using river water for public water supplies, the pesticide concentration patterns in finished tap water are very similar to those of the rivers themselves (Baker, 1983). Either granular activated carbon or powdered activated carbon can be used to reduce pesticide concentrations in the river water.* At the present time, federal drinking water standards have not been established for any of the compounds listed in Tables 7 and 8. However, it is expected that standards will be established for some of them in the near future.

Pesticides in Rainwater

Several currently-used pesticides are also present in rainwater. Table 10 illustrates the pesticide concentrations in rainfall samples collected in Tiffin, Ohio in 1985. Similar pesticide concentrations were found in rainwater samples collected in West Lafayette, Indiana, while samples collected in Potsdam, New York and Parsons, West Virginia had much lower concentrations. While the pesticide concentrations in rainwater are relatively high relative to other synthetic organic compounds in rainwater or in comparison with previous reports of pesticides in rainwater (Tabatabai, 1983), the total pesticide deposition with rainfall is much less than application rates of these same compounds. Consequently, direct herbicidal action on either target or nontarget plants from herbicides in rainwater would not be expected (Richards et al., 1986).

CONCLUSIONS

The tributary monitoring programs in the Lake Erie Basin indicate that area rivers transport large loads of sediments and both particulate and soluble phosphorus derived from agricultural sources. Agricultural runoff also results in riverine nitrate concentrations which frequently exceed drinking water standards. Relatively high concentrations of many currently used pesticides are also present in area streams and public water supplies during runoff events following pesticide application.

---

*Research on pesticide removal techniques for municipal water supplies is currently underway through a cooperative agreement between Heidelberg College and the U.S. EPA, Drinking Water Research Division, Cincinnati Ohio.

Table 9. Time Weighted Average Concentrations (µg/l) of Pesticides for the April 15 Through August 15 Periods for 1983, 1984 and 1985 at Three River Transport Stations in Northwest Ohio. Calculated with 7 Days as Maximum Duration for Any Sample to Characterize the Stream Concentration

	Honey Creek			Sandusky River			Maumee River		
	1983	1984	1985	1983	1984	1985	1983	1984	1985
N	57	72	88	45	53	62	43	58	38
Simazine	nd	0.05	0.17	nd	0.15	0.20	nd	0.18	0.18
Carbofuran	0.11	0.27	0.26	0.16	0.14	0.19	0.18	0.19	0.04
Atrazine	3.07	4.46	5.29	1.82	2.53	4.43	1.77	2.98	2.05
Terbufos	<0.001	nd	0.003	nd	nd	0.001	0.001	<0.001	0.001
Fonofos	nd	nd	0.001	0.004	nd	0.004	nd	0.002	<0.001
Metribuzin	0.36	0.27	0.66	0.30	0.37	0.92	0.45	0.45	0.28
Alachlor	1.41	2.12	2.13	0.51	1.25	1.88	1.06	1.76	0.52
Linuron	0.34	0.05	0.67	0.09	0.003	0.33	nd	0.04	0.01
Metolachlor	3.05	3.00	4.41	2.27	2.72	4.84	0.26	1.57	1.44
Cyanazine	0.67	0.65	1.16	0.45	0.49	0.62	0.56	1.14	0.35

Table 10. Pesticide Concentrations (μg/l) in Rainwater Collected at Tiffin, Ohio in 1985

Date	Butylate	Carbofuran	Atraxine	Fonofos	Alachlor	Metolachlor	Cyanazime
04/05							
04/07							
05/01	0.226		1.043	0.039	0.822		0.351
05/15			1.044	0.242	3.702	1.821	0.127
05/16			0.581		3.162	1.094	0.041
05/16	0.063		0.210	2.009	0.525	0.022	
05/17			0.213	1.752		0.046	
05/20		0.110	0.685	0.942		0.134	
05/27			0.203	0.507	0.141	0.057	
10/04			0.146	0.267			
06/09			0.295	0.501	0.037	0.036	
16/11			0.342	0.599	0.273	0.028	
16/12			0.103	0.204			
16/15			0.068	0.271	0.274		
06/22			0.138	0.193	0.111	0.028	

Much discussion during this workshop has focused on the issues of whether various forms of conservation tillage would increase nitrate and pesticide contamination in surface water and ground water. Given the extent of nitrate and pesticide contamination associated with current management practices in this region, our concerns should be not only that conservation tillage not aggravate these problems, but that comprehensive agricultural management programs be developed that will reduce nitrate, pesticide, and soluble phosphorus problems, as well as sediment and particulate phosphorus loads. Residue management must be accompanied by improved fertilizer and pesticide management if we are to most efficiently address agricultural nonpoint source pollution problems in the Great Lakes Region. The emphasis within the Lake Erie demonstration programs has been shifting toward this more comprehensive approach to agricultural land management. This approach should not only reduce agricultural nonpoint pollution but should also improve the economic returns for area farmers.

The Lake Erie tributary monitoring programs also illustrate the substantial effects of watershed size on the characteristics of agricultural pollution in stream and river systems. These studies suggest that as watershed size becomes smaller:

- annual variability in material export increases;
- peak concentrations of sediment, nutrients and pesticides increase;
- the duration of intermediate concentrations of sediments, nutrients and pesticides decreases;
- smaller proportions of time account for larger proportions of total material export;
- more samples must be collected and analyzed in order to accurately measure material export;
- the seasonal distribution of stream sediment transport corresponds more closely to the seasonal distribution of erosion processes.

The effects noted above are a consequence of several factors. As watershed size increases, the role of material transport and processing within stream systems becomes more significant. Also as watershed size increases, both weather-related and management-related inputs tend to average out and show less year-to-year variability. An awareness of these watershed "scale" effects is important in establishing watershed monitoring programs and in interpreting and comparing data on the extent of agricultural pollution based on stream monitoring programs.

REFERENCES

Adams, J. R., T. J. Logan, T. H. Cahill, D. R. Urban, and S. M. Yaksich. 1982. A land resource information system (LRIS) for water quality management in the Lake Erie Basin. J. Soil and Water Cons. 37(1): 45-50.

Anonymous. 1985. Lake Erie conservation tillage demonstration projects: Evaluating management of pesticides, fertilizer, residue to improve water quality. Conservation Tillage Information Center, Fort Wayne, Indiana, 1985. 20 pp.

Baker, D. B. 1980. Upstream point source phosphorus inputs and effects. Seminar on Water Quality Management Trade-Offs (Point Source vs. Diffuse Source Pollution) Conference, September 16-17, 1980. U.S. EPA, Chicago, ILL., EPA-905/9-80-009, pp. 227-239.

Baker, D. B. 1983. Herbicide contamination in municipal water supplies of northwestern Ohio. Draft final report to the Joyce Foundation, Chicago, IL. Heidelberg College, Tiffin, OH. 33 pp. + appendix.

Baker, D. B. 1984. Fluvial transport and processing of sediments and nutrients in large agricultural river basins. EPA-600/8-83-054. U.S. Environmental Protection Agency, Athens, GA. 168 pp.

Baker, D. B. 1985. Regional water quality impacts of intensive row-crop agriculture: A Lake Erie Basin case study. Journal of Soil and Water Conservation, 40(1):125-132.

Crosson, Pierre. 1981. Conservation tillage and conventional tillage: A comparative assessment. Soil Cons. Soc. Am., Ankeny, Iowa.

Great Lakes Phosphorus Task Force. 1985. United States task force plan for phosphorus load reductions from non-point, and point sources on Lake Erie, Lake Ontario, and Saginaw Bay. Great Lakes National Program Office U.S. EPA Chicago, Illinois, 1985. 176 pp.

Hinkle, M. K. 1983. Problems with conservation tillage. Journal of Soil and Water Conservation, 38(3): 201-206.

Honey Creek Joint Board of Supervisors. 1982. Honey Creek watershed project, 1979-1981. U.S. Army Corps of Engineers, Buffalo District, Buffalo, New York, 1982.

International Joint Commission. 1978. Environmental management strategy for the Great Lakes system. Final report from PLUARG. Windsor, Ontario. 115 pp.

International Joint Commission. 1980. Pollution in the Great Lakes Basin from land use activities. Report to the governments of the United States and Canada. International Joint Commission. Windsor, Ontario. 141 pp.

International Joint Commission. 1983. Nonpoint source pollution abatement in the Great Lakes Basin: An overview of post-PLUARG developments. Windsor, Ontario, 129 pp.

Kramer, J. W. and D. B. Baker. 1985. An analytical method and quality control program for studies of currently used pesticides in surface waters. In: Taylor, J. K. and T. W. Stanley, eds. Quality assurance for environmental measurements, ASTM STP 867. Amer. Soc. Testing and Materials, Philadelphia. pp. 116-132.

Logan, T. J., D. R. Urban, J. R. Adams, and S. M. Yaksich. 1982. Erosion control pontential with conservation tillage in the Lake Erie Basin: Estimates using the universal soil loss equation and the Land Resource Information System (LRIS). J. Soil and Water Con. 37(1):50-55.

Macknis, J. 1985. Chesapeake Bay nonpoint source pollution. U.S. EPA Perspectives on nonpoint source pollution. Proceedings of a national conference, May 19-22, 1985, Kansas City, Missouri. EPA 440/5-85-001, pp. 165-171.

Morrison, J. B. 1984. Lake Erie demonstration projects evaluating impacts of conservation tillage on yield, cost, environment. National Association of Conservation Districts, 17 pp.

Ohio EPA. 1985. State of Ohio phosphorus reduction strategy for Lake Erie. Ohio Environmental Protection Agency, Columbus, Ohio. 89 pp. + appendices.

Rast. W. and G. F. Lee. 1983. Nutrient loading estimates for lakes. J. Environ. Eng. 109(2):502-517.

Resource Management Associates. 1979. Land resources information for the Lake Erie drainage basin. Lake Erie Wastewater Management Study, U.S. Army Engineer District, Buffalo, NY. 63 pp.

Richards, R. P., J. W. Kramer, D.B. Baker, and K. A. Krieger. 1986. Agricultural herbicides and insecticides in rainwater in the northeastern United States. [Internal report, Water Quality Laboratory, Heidelberg College, Tiffin, Ohio]. 15 pp.

Smullen, J. T., J. L. Taft, and J. Mackins. 1982. Nutrient and sediment loads to the tidal Chesapeake Bay system. Chesapeake Bay program technical studies: A synthesis. U.S. EPA, Washington, D.C. pp. 150-265.

Tabatabai, M. A. 1983. Atmospheric deposition of nutrients and pesticides. In Agricultural management and water quality, edited by Frank W. Schaller and George W. Bailey. Iowa State U. pr., Ames, IA. pp. 92-108.

U.S. Army Corps of Engineers, Buffalo District. 1982. Lake Erie wastewater management study. Final report. Buffalo, NY. 225 pp.

Verhoff, F. H., D. A. Melfi, S. M. Yaksich, and D. B. Baker. 1978. Phosphorus transport in rivers. Technical report series. Prepared for the Lake Erie Wastewater Management Study, U.S. Army Engineer District, Buffalo, NY. 88 pp.

SECTION II

EFFECT OF CONSERVATION TILLAGE SYSTEMS ON SOIL PHYSICAL, CHEMICAL AND BIOLOGICAL PROCESSES

CHAPTER 5

HYDROLOGIC SOIL PARAMETERS AFFECTED BY TILLAGE

C. A. Onstad and W. B. Voorhees,
USDA-ARS, North Central Soil Conservation
Research Laboratory, Morris, Minnesota

INTRODUCTION

Conservation tillage as defined in the Resource Conservation Glossary (SCSA, 1982) is "Any tillage system that reduces loss of soil or water relative to conventional tillage." The literature abounds with information on the effects of conservation tillage on hydrology and soil physical conditions. However, much of the information is generic and qualitative, sometimes resulting in seemingly conflicting results. Usually, apparent conflicting information results from incomplete data regarding location, history, and timing of supposedly identical experiments.

The purpose here is to examine conservation-tillage induced changes in soil physical conditions affecting hydrologic response. Conservation tillage leaves considerable residue on the soil surface, leaves the soil surface rough or ridged, and mixes the residue and soil to varying degrees within the management zone. After tillage, the presence of residue together with soil physical parameters and weather influence the degradation rate of the conditions established. Weather combined with soil physical properties also determine the alleviation rate of adverse soil conditions caused by wheel track compaction. These charactersitics will be discussed with respect to their affects on hydrologic processes.

Effects of Conservation Tillage on Groundwater Quality: Nitrates and Pesticides, Terry J. Logan et al., eds.

Residue Management

Residue management practices have been grouped into four classes by Van Doren and Allmaras (1978). These classes include:

1. Residues at or above the soil surface.
2. Residues mixed into the soil surface.
3. Residues incorporated into the tilled layer.
4. Residues completely removed.

Residues in each of these classes affect hydrology and soil physical properties differently.

Residues at the surface protect the soil surface from both radiation and rainfall energy. Such residue will slow the rate of soil cooling and warming. In addition, evaporation may be reduced from residue covered soils. Surface sealing is partly the result of rain-rainfall energy striking the soil surface, thus rainfall interception by residue retards the formation of a surface seal. Overall, the effects of residue on the soil are to maintain the ambient soil temperature, reduce evaporation, and maintain infiltration capacity.

When residues intercept the soil surface or are anchored to the soil, they provide additional protection for the soil surface. First, they provide a conduit during saturated conditions through which water can be conducted into the tilled layer under saturated conditions. Anchored residues also provide opportunity for saturated flow through cracks surrounding plant roots. Additionally, anchored roots provide increased resistance to flowing water increasing the opportunity for infiltration and decreasing the shear stress thereby resulting in less soil detachment. Partially standing residues left over-winter trap additional water in the form of snow.

Residues partially or completely incorporated into the tilled zone also have a long-term effect on soil physical properties and hydrology. Residue decomposition increases soil tilth by increasing organic matter, improving soil structure, and forming more stable aggregates with respect to the forces of raindrop impact and flowing water. Complete residue removal exposes the soil surface to the full force of the weather.

Soil Surface

The soil surface is a very dynamic, critically important, hydrologic interface. Soil surface condition can be described in terms of roughness, strength, stability, porosity, pore-size distribution, and conductivity. All of these factors are influenced by basic soil characteristics interacting with forces due to tillage, weather, and wheel traffic.

At times, some soils appear quite smooth. However, most soils contain small depressions or roughness referred to as microrelief. Soil roughness is the result of tillage and has considerable impact on hydrologic processes. Rougher soils frequently maintain higher infiltration rates because dense crusts tend to form mainly in surface depressions (Larson, 1962). Rough soils are generally more porous than smooth soils resulting in larger total water holding capacities (Voorhees et al., 1981). Primary tillage increases roughness and porosity, while secondary tillage usually decreases roughness and tilled layer porosity as shown in Table 1 (Voorhees, et al., 1981).

The roughness and stability of the surface soil structure is affected not only by tillage but also by water content at time of tillage and presence or absence of wheel traffic compaction before tillage. Allmaras et al. (1966) reported that the surface structure produced by a given tillage operation on a given soil type will depend on the water content at time of tillage. Voorhees et al. (1979) reported that the random roughness of a clay loam soil was 24.1 mm after fall moldboard plowing if the soil had not been compacted during the growing season. Where there had been seasonal wheel traffic, the random roughness after plowing increased to 37.3 mm. This increased roughness increases the potential to trap and store water.

Table 1. Random roughness and fractional total porosity of tilled layer as affected by tillage systems (Voorhees et al., 1981)

Tillage method		Random roughness measured after spring tillage (cm)		Total fractional porosity measured after spring tillage	
Fall	Spring	Range	Mean	Range	Mean
Plow	--	1.96-7.26	3.24	0.54-0.96	0.72
Tandem disk	--	1.75-1.83	1.79	0.54-0.69	0.61
Chisel	--	1.35-1.68	1.52	0.68-0.69	0.68
Plow	--	1.30-3.63	2.22	0.48-0.68	0.60
Plow	Disk and harrow	0.89-1.52	1.27	0.58-0.66	0.61
None	Plow	1.17-3.61	2.28	0.55-0.84	0.66
	Plow, disk and harrow	0.69-2.16	1.15	0.53-0.82	0.67
None	Plow and wheel tracked	0.93-1.15	1.06	0.52-0.56	0.54
None	Rotary till	1.55-1.75	1.65	0.64-0.67	0.66
None	None	0.46-0.86	0.59	0.52-0.53	0.52

Larson (1964) included microrelief as a critical factor for evaluating tillage requirements for corn. Rough soils temporarily store more water in depressions than do smooth soils. Increased depressional storage traps more sediment, reducing erosion and increasing infiltration. Onstad (1984) developed a relationship among depressional storage, slope, and random roughness as shown in Figure 1. This figure indicates that depressional storage ranges up to about 1 cm for relatively rough, flat soils. The value of 1 cm may be deceptively low because, during a storm, filled depressions may infiltrate and refill several times as rainfall intensities decrease to below infiltration capacity and then increase again.

As surface depressions fill during the rainfall, the air-water interface area increases. The fraction of the area covered with water varies with roughness and slope steepness (Fig. 2). The amount of surface area covered with water is hydrologically important for two reasons. Soil covered with water at depths above 8 mm is protected from the forces of raindrop impact and probably does not erode. Instead, depressions are sediment traps retaining soil eroded elsewhere. Conversely, areas of soil exposed to rainfall provide information on sources of erosion of tilled surfaces subject to raindrop impact.

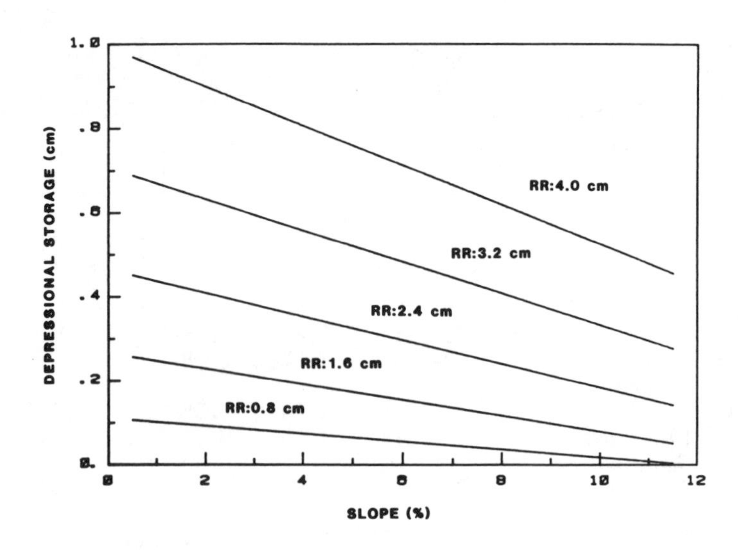

Figure 1. Maximum depression storage as a function of random roughness and slope steepness (Onstad, 1984).

Figure 2 indicates that, for flat slopes, depressional surface area increases steadily up to 50 percent of maximum as soils become rough. For steeper slopes, depressional areas increase and then begin to decrease as random roughness increases indicating that depressions become more interconnected as roughness increases.

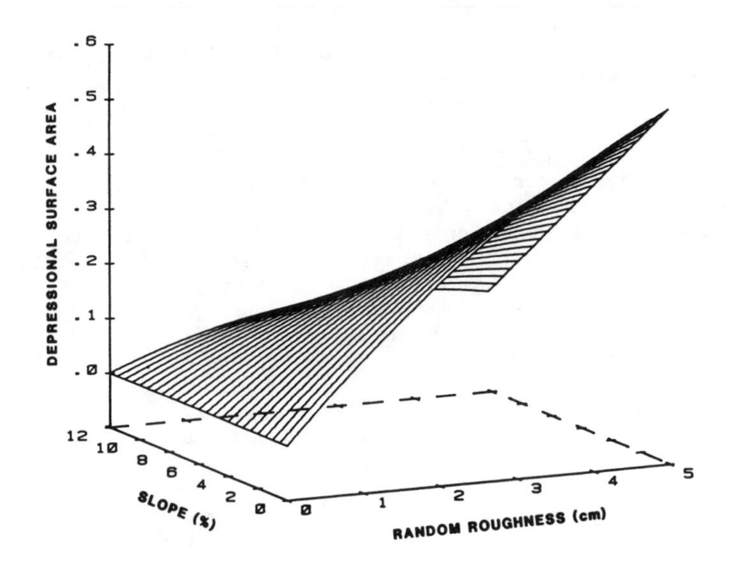

Figure 2. Ratio of surface area of depressions to total surface area as a function of random roughness and slope steepness (Onstad, 1984).

After rainfall begins, the hydraulic conductivity of the soil surface becomes the limiting factor for infiltration in many soils (Van Doren and Allmaras, 1978). The conductivity of the surface layer decreases as surface seals form. Surface seals can form both in depressions and on exposed surfaces. Sediment accumulates in depressions forming a crust. On exposed surfaces, rainfall energy impacts directly on the soil surface causing two processes to begin. First, soil particles are rearranged to form a more dense matrix. For soils having little or no aggregation, rearrangement is rapid, quickly causing reductions in hydraulic conductivity (Bosch, 1986). Second, for well aggregated soils, both individual aggregates and primary soil particles are packed into a more dense matrix. In addition, soil aggregates are broken down by raindrop energy into primary particles which contribute to the sealing phenomena (Wolfe, 1986). The decrease in hydraulic conductivity after disturbance is shown in Figure 3, taken from Bosch (1986), for four soils in western Minnesota. The addition of water to tilled soils in the absence of kinetic energy also reduces random roughness and hydraulic conductivity (Onstad, et al., 1984) but to a lesser extent than when rainfall kinetic energy is applied.

Soil strength is related to many physical and chemical soil properties. Generally, soil texture is related to strength with silt and clay contents being most important (Mannering, 1967). As fine particles increase, specific surface area increases contributing to more and stronger bonds (Kemper et al., 1974). Clay type also is related to aggregate strength (Lemos and Lutz, 1957). In addition, water content, organic matter content, cation exchange capacity, and rate of wetting or drying are related to soil and aggregate strength.

Surface soil characteristics affected by tillage are also affected by wheel traffic compaction. Wheel traffic can both negate and enhance tillage effects. The most obvious effect of wheel traffic is to cause surface depression. Depending on soil conditions and the load being carried by the wheel, the depression can easily be 10 cm in depth (Davies et al., 1973).

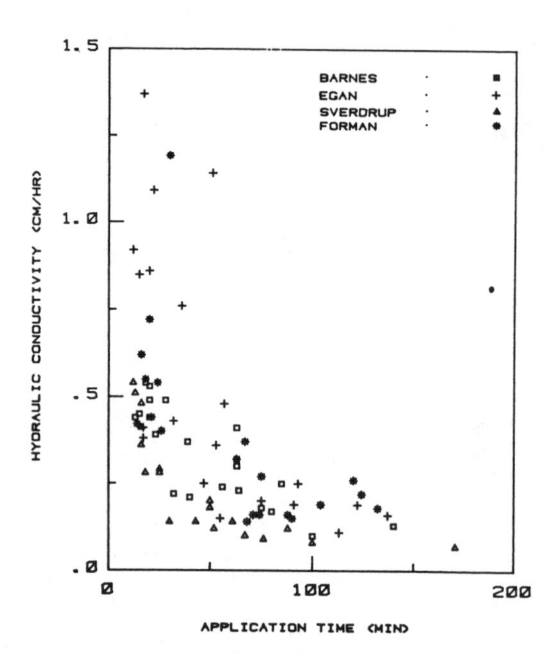

Figure 3. Hydraulic conductivity at the time of ponding versus application time for four soils (Bosch, 1986).

Thus, the increase in surface depressional storage due to rough primary tillage can easily and quickly be negated by a single passing of a wheel. On the other hand, a relatively smooth, loose surface condition created by secondary tillage, and which may have relatively low surface depressional storage capacity, can theoretically have its depressional storage capacity significantly increased by subsequent wheel traffic. Depending on orientation with respect to slope, these ruts may temporarily store surface water or quickly shed surface water.

Characteristics of surface soil structural units themselves are also altered by wheel traffic. Voorhees et al., (1979) showed that density of individual soil clods about 5 cm in diameter was increased from 1.47 to 1.73 Mg/m^3 by one season's tractor wheel traffic. The crushing strength of these compacted clods was four times higher than for noncompacted clods. The stability of wheel-trafficked soil geo-structural units persisted over winter (Voorhees, 1977), with the metric mean diameter the following spring being about 40 mm compared to only 7 mm for soil that had not been wheel trafficked (Voorhees et al., 1978). Thus, during the off-season (from primary fall tillage until secondary spring tillage) wheel traffic can induce an additional degree of stability to the surface soil structure that is beneficial for trapping and temporary storage of water. The net result should be less water runoff. These results were from clay loam soils, relatively high in organic matter. Sandy soils or soils lower in organic matter might not exhibit the same degree of change.

Wheel traffic associated with spring planting of row crops will usually have negative effects on soil surface conditions with respect to hydrology. Decreased infiltration and increased runoff from wheel-tracked soil compared to nontracked soil (measured after planting) has been documented for a number of soil textures (Young and Voorhees, 1982; Lindstrom and Voorhees, 1980). Instability of the surface, resulting in early sealing, was a common factor in these studies.

Soil Management Zone

Tillage

By definition, "soil management zone" infers a layer of soil in which man has some capability of managing. With respect to hydrology, the objective of the management effort is to control the movement of water (and associated material) within the soil profile once the water has crossed the soil surface. Historically, tillage has been the most effective way of changing soil physical properties, and hence the soil management zone has come to mean the tillage zone. Due to a wide range of soil and climatic conditions, plus specific crop needs, the depth of this soil management zone is not constant but normally includes approximately the surface 20 cm. Recent changes in agriculture have both decreased and increased soil management zone thickness.

The soil management zone becomes thinner and closer to the surface as tillage intensity decreases. An offsetting trend, however, is compaction from wheel traffic that tends to increase the depth of the management zone. A need for occasional deep tillage is often associated with subsoil compaction. Tillage generally, but not always, has a net loosening effect, while wheel traffic has a consolidating effect. Both of these factors affect hydrology by influencing the water storage and transmission characteristics of the management zone.

Primary tillage initially increases total pore volume of the tilled layer. Compared to an untilled soil with an initial porosity of 52%, moldboard plowing increased total pore volume in the surface 15 cm to 72%; disking increased it to 61% (Voorhees et al., 1981). This results in an increased potential tilled layer water storage capacity of 3.0 and 1.35 cm for the plow and disk treatments, respectively. Subsequent wheel traffic can easily negate these tillage effects (Voorhees, et al., 1978). Antecedent soil water content will affect the change in porosity of a given soil in response to a given tillage operation. Allmaras et al. (1966) reported that secondary tillage increased total porosity when the soil moisture ratio (gravimetric water content at tillage divided gravimetric water content at the lower plastic limit) approached a value of 1.0. Secondary tillage tended to decrease total porosity if this ratio was less than 1.0 (somewhat drier).

Rawls et al. (1983) used published data to stratify soil porosity according to soil texture after moldboard plowing. Their results (Figure 4) show that finer textured soils increase porosity less than coarse textured soils. Porosity decreases as a result of subsequent tillage practices and natural processes during the growing season. Parenthetical values shown in Figure 4 indicate the average seasonal decrease of porosity.

Total porosity is important to total water storage capacity and water flow. Warkentin (1971) reviewed several data sets and reported that the log of saturated hydraulic conductivity increased linearly as void ratio (total voids volume divided by total solids volume) increased. Void ratio changes, caused by tillage or wheel traffic compaction, are most commonly associated with macropore changes.

Macropores are either created by tillage or destroyed by compaction or excessive tillage, with micropores being subject to less change. Saturated hydraulic conductivity is very sensitive to the quantity of continuous large pores in the soil system.

Total pore space was reported to be an important factor determining infiltration rates in tillage experiments on a clay loam soil (Lindstrom and Voorhees, 1980; Lindstrom et al., 1981). In these studies, no-till generally had a lower total porosity than did conventionally tilled soil. In other studies, large macropores attributed to earthworm activity resulted in no-till having significantly higher saturated conductivity and higher infiltration rates than tilled treatments (Edwards and Norton, 1985; Lal, 1978). Thus, supposedly similar tillage treatments should not be expected to produce similar results over a range of soil textures.

Most movements of water in soils during infiltration and redistribution involves unsaturated flow, and is sensitive to soil water content and gradients of soil water potential. Hanks and Ashcroft (1980) reviewed the general theory regarding infiltration for two cases. In the first case (Fig. 5), water is being applied at the soil surface at a rate exceeding saturated hydraulic conductivity.

The point in time when infiltration rate decreases corresponds to initiation of water ponding on the surface, with potential runoff resulting. The time required to reach that point is a function of the volumetric water content of the surface layer. If the mean pore diameter of the surface layer is decreased (either by wheel traffic compaction or through lack of tillage), the volumetric water content of that layer will reach saturation sooner; thus runoff will potentially occur sooner when soil is in a compacted state. Also, a compacted soil will have a lower volumetric water content at saturation (Voorhees et al., 1979) and thus ponding and potential runoff will occur sooner than for a noncompacted soil.

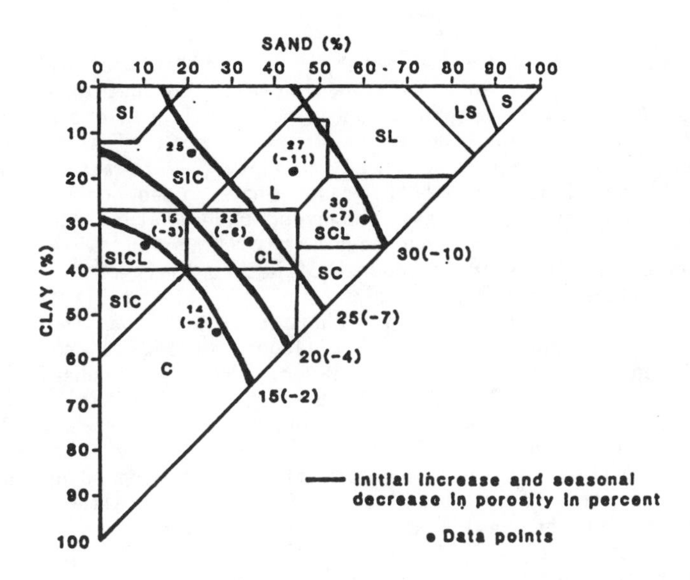

Figure 4. Percent porosity increases due to a moldboard plow (percent decreases during the growing season) as a function of soil texture (Rawls, et al., 1983).

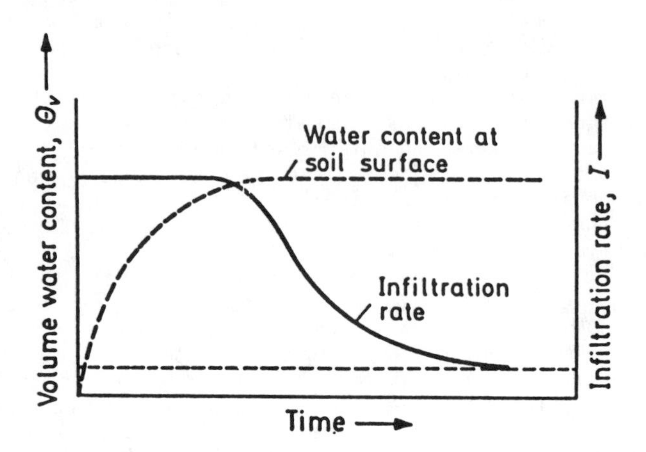

Figure 5. Vertical water infiltration into soil.

For the second case, water is applied at the soil surface at a rate less than saturated hydraulic conductivity. Water movement is now in response to the potential gradient within the profile as well as the water content, both of which can be altered by tillage and wheel traffic. Mathematical models describing flow for both cases have been developed for homogeneous soil profiles. However, under natural conditions, soil profiles are heterogeneous with tillage and wheel traffic compaction increasing the degree of heterogeneity.

Tillage normally results in the tilled layer having distinctly different pore size characteristics than the untilled layer. Depending on the tillage method, there may be differences within the tilled layer. Subsequent wheel traffic tends to restore the tilled layer to its initial state of porosity and pore size distribution (Lindstrom et al., 1981). In addition, wheel traffic may go beyond simply "restoring" a tilled soil to its previous untilled state--depending on soil water content and load applied, it may alter porosity characteristics considerably deeper than normal tillage depth (Ericksson et al., 1974; Sommer, 1976; Voorhees et al., 1986). The various combinations of wheel traffic compaction and tillage produce an endless combination of soil layers, each with its own unique hydrologic characteristics. Further complications arise because soil layers are often not well defined but form a continuum. Allmaras et al. (1982) showed that a compacted tillage pan only a few cm thick can have a profound effect on water movement through the soil profile.

An effect of tillage is to alter the amount, orientation, and anchorage of plant residue within the surface layer of soil (Gebhardt, et al., 1985). A concentrated layer of residue incorporated at some depth can have the same effect on water redistribution as can a layer of highly permeable soil overlying a less permeable soil. Under saturated conditions, the residue mat will not retard water movement, but can act as a non-permeable layer if flow is under unsaturated conditions.

Wheel traffic can also introduce a horizontal component of variability with respect to water movement. During a drying cycle, wheel traffic compacted zones may impose a higher soil water potential gradient within the profile but, because of its low conductivity, may actually retard net water loss (Reicosky et al., 1981). The opposite may be true for a wetting cycle under unsaturated conditions (Kemper et al., 1971).

Finally, it must be realized that conservation tillage by itself does not automatically eliminate compaction related concerns. Regardless of the type of tillage used, there will always be harvest equipment involved. Recent field studies are confirming laboratory theory that subsoil compaction is a function of total axle load (Voorhees et al., 1986; Taylor et al., 1980). Harvest equipment is generally the heaviest equipment used on modern farms. In spite of annual freezing and thawing, subsoil compaction from such equipment can persist across seasons and affect water movement through the soil profile (Voorhees et al., 1986) and soil erosion (Fullen, 1985).

Natural Degradation

Principal natural causes changing soil physical properties in the tilled layer are kinetic energy of rainfall, wetting-drying cycles, and freezing-thawing cycles. These processes produce measurable effects on total pore space, pore-size distribution, bulk density, and hydraulic conductivity. Kemper and Koch (1966) found that aggregate stability increased with increases in organic matter, clay content, and nitrogen content and decreased with increased exchangeable sodium percentage.

It is usually difficult to separate rainfall energy effects from wetting-drying action in the soil management zone as both are occurring simultaneously. Onstad et al. (1984) used a fiberglass furnace filter to eliminate raindrop splash action when using a rainfall simulator to study effects of water application only on a freshly tilled surface. They found that adding 15 cm of water increased bulk density up to 20% and decreased saturated hydraulic conductivity as much as 74%. Kinetic energy due to rainfall is dissipated through the destruction and rearrangement of soil aggregates and, thus, affects the surface layer more than the remainder of the soil management zone.

Burwell et al. (1966) found that total pore space decreased from 13.7 cm to 11.2 cm for a freshly plowed soil after 5 cm of runoff had occurred from rainfall applied artificially at 12.5 cm/hr. They also stated that 80% of the decrease in total pore space occurred prior to initiation of runoff. In addition, they found that total pore space accounted for 60% of the variation required to initiate runoff.

Freezing and thawing usually destroy soil structure and aggregation (Bisal and Nielsen, 1964) although structural improvements have been observed in some soils (Sillanpaa and Webber, 1961). The effect of freezing and thawing cycles on soils is dependent on size of soil aggregates, antecedent soil water content, and level of ambient freezing temperature. Benoit and Bornstein (1970) found that 15 freeze-thaw cycles increased the soil water content in soil aggregates by an average of 38% over initial values. Soil water content in the largest aggregates (up to 19 mm) increased more than for smaller aggregates (less than 2 mm) regardless of the freezing temperature. This was true for the aggregates at maximum water holding capacity, implying that pore-size distribution changed in a manner such that more water is retained by the soil. This change was likely a shift to smaller pores which do not drain so freely.

In another study, Benoit (1973) found that large aggregates decreased after 20 freeze-thaw cycles. Here the soil water content at the time of freezing, the ambient freezing temperature, and the initial aggregate size were instrumental in determining the degree of reduction. Table 2 from Benoit (1973) shows these relations. The table shows also that, initially, small aggregates increased in size after freezing-thawing particles at lower antecedent soil water levels and higher freezing temperatures.

Freeze-thaw cycling also affects soil bulk density depending on initial aggregate size, soil water content, and temperature. Largest increases in bulk density occur with large, wet aggregates frozen to very low temperatures. Changes for other combinations are shown in Table 3 after 15 freeze-thaw cycles (Benoit and Bornstein, 1970).

Changes in hydraulic conductivity after freeze-thaw cycles are more complex. When aggregates are dry, hydraulic conductivity increases after freeze-thaw cycles with the opposite occurring for wetter aggregates. Figure 6 from Benoit (1973) shows the magnitude of these changes. The figure indicates that hydraulic conductivity is increased five times for dry aggregates, whereas for wet aggregates, the final conductivity is only 0.1 of that initially.

Table 2. Ratio of final to initial percentage of water-stable aggregates retained on a 0.50-mm screen

Aggregate size group[3]	Water treatment level[2]: Maximum water holding capacity, Freezing temp. °C: -18	-4	0.5 bar pressure, Freezing temp. °C: -18	-4	Average
0.0-0.8	0.92[1]	1.07	1.10	1.26	1.09
0.8-1.2	0.72	0.74	0.98	1.09	0.88
1.2-2.0	0.71	0.84	0.89	0.78	0.83
Average	0.78	0.92	0.99	1.04	

1. Each value is the average for three cores with duplicate determinations run on each core from Benoit (1973).
2. LSD 0.05 between overall water level means = 0.14.
3. LSD 0.05 between overall aggregate mean values = 0.17.

Table 3. Relative change in bulk density after 15 freeze-thaw cycles as related to aggregate size, water content and freezing temperature (ratio of initial to final) (Benoit and Bornstein, 1979).

Aggregate size, mm	Water treatment level: Maximum water holding capacity, Freezing temp. °C: -18	-2	0.5 bar pressure, Freezing temp. °C: -18	-4	Average
0.00-2.00	0.92	0.86	0.93	0.94	0.90
2.00-4.8	0.75	0.77	0.83	0.84	0.80
4.8 - 19.1	0.73	0.77	0.79	0.80	0.77
Average	0.78	0.80	0.85	0.86	

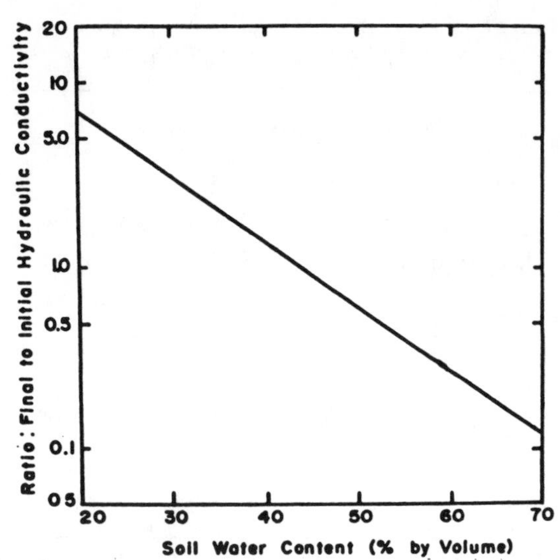

Figure 6. Semilog plot of the ratio of final to initial soil hydraulic conductivity after freezing and thawing soil cores at different soil water contents (Benoit, 1973).

SUMMARY

Tillage affects soil properties which in turn affect hydrologic response. Often, studies are conducted relating runoff rates and quantities of rainfall, both expressed precisely, without regard for quantitative expression of the soil characteristics. In order for trends to be extrapolated, it is not sufficient to use qualitative terms such as "moldboard plowing" or "no-till" to express soil conditions during the event. Instead, the soil needs to be quantified in terms of residue cover, surface roughness, structure, aggregate stability, total pore space, and pore size distribution.

Tilled soil physical properties are dynamic. They are affected by type of tillage, tillage sequence, wheel traffic, and natural degradation factors. Most important, natural factors include rainfall kinetic energy, soil wetting and drying cycles, and soil freezing and thawing cycles.

Included are descriptions of the major affected soil physical properties. This includes the effects of residue management, soil surface characteristics, and the soil management zone as influenced by tillage and natural degradation. These are related to tillage characteristics and natural degradation.

REFERENCES

Allmaras, R. R., R. E. Burwell, W. E. Larson, and R. F. Holt. 1966. Total porosity and random roughness of the interrow zone as influenced by tillage. USDA Conservation Research Report No. 7.

Allmaras, R. R., K. Ward, C. L. Douglas, Jr., and L. G. Ekin. 1982. Long-term cultivation effects on hydraulic properties of a Walla Walla silt loam. J. Soil and Tillage Res. 2(3):265-280.

Benoit, G. R. and J. Bornstein. 1970. Freezing and thawing effects on drainage. Soil Sci. Soc. Am. Proc. 34:4:551-557.

Benoit, G. R. 1973. Effect of freeze-thaw cycles in aggregate stability and hydraulic conductivity of three soil aggregate sizes. Soil Sci. Soc. Am. Proc. 37:1:3-5.

Bisal, F. and K. F. Nielsen. 1964. Soil aggregates do not necessarily break down over winter. Soil Sci. 98:345.

Bosch. D. D. 1986. The effects of rainfall on the hydraulic conductivity of soil surfaces. Unpublished M.S. Thesis. University of Minnesota. 161 pp.

Burwell, R. E., R.R. Allmaras, and L. L. Sloneker. 1966. Structural alteration of soil surfaces by tillage and rainfall. J. Soil and Water Conserv. 21:2:61-63.

Davies, D. B., J. B. Finney, and S. J. Richardson. 1973. Relative effects of tractor weight and wheel-slip in causing soil compaction. J. Soil Sci. 24(3):399-409.

Edwards, W. M. and L. D. Norton. 1985. Characterizing macropores after 20 years of continuous no-tillage corn. Agronomy Abst. p. 205.

Ericksson, J., I. Hakansson, and B. Danfors. 1974. The effect of soil compaction on soil structure and crop yields. Swedish Institute of Agricultural Engineering Bulletin 354. Uppsala, Sweden.

Fullen, M. A. 1985. Compaction, hydrological processes and soil erosion on loamy sands in east Shropshire, England. J. Soil and Tillage Research, 6(1):17-30.

Gebhardt, M. R., Daniel, T. C., Schweizer, E. E., and Allmaras, R. R. 1985. Conservation Tillage. Science 230:625-630.

Hanks, R. J. and G. L. Ashcroft. 1980. Applied soil physics: Soil water and temperature applications. Advanced Series in Agricultural Sciences No. 8. Springer-Verlag, New York.

Kemper, W. D. and E. J. Koch. 1966. Aggregate stability of soils from western United States and Canada. USDA Tech. Bull. No. 1355. 52 pp.

Kemper, W. D., D. D. Evans, and H. W. Hough. 1974. Crust strength and cracking. In John W. Cary and D. D. Evans (eds.) Soil Crusts. Arizona Agric. Expt. Stn. Tech. Bull. 214, p. 31-39.

Kemper, W. D., B. A. Stewart, and L. K. Porter. 1971. Effects of compaction on soil nutrient status. pp. 178-189. In K. K. Barnes, W. M. Carleton, T. M. Taylor, R.I. Throckmorton, and G.E. Vanden Berg (eds.) Compaction of Agricultural Soils. ASAE Monograph, St. Joseph, MI.

Lal, R. 1978. Influence of tillage methods and residue mulches on soil structure and infiltration rate. pp. 393-402. In W.W. Emerson, R. D. Bond and A. R. Dexter (eds.) Modification of soil structure. John Wiley & Sons, New York.

Larson, W. E. 1962. Tillage requirements for corn. J. Soil Water Conserv. 17:3-7.

Larson, W. E. 1964. Soil parameters for evaluating tillage needs and operations. Soil Sci. Soc. Am. Proc., 28:119-122.

Lemos, P. and J. F. Lutz. 1957. Soil crusting and some factors affecting it. Soil Sci. Soc. Am. Proc. 21:485-491.

Lindstrom, M. J. and W. B. Voorhees. 1980. Planting wheel traffic effects on interrow runoff and infiltration. Soil Sci. Soc. Am. J. 44:84-88.

Lindstrom, M. J., W. B.Voorhees, and G. W. Randall. 1981. Long-term tillage effects on interrow runoff infiltration. Soil Sci. Soc. Am. J. 45:945-948.

Mannering, J. V. 1967. The relationship of some physical and chemical properties of soils to surface sealing. Ph.D. Thesis, Purdue University.

Onstad, C. A. 1984. Depressional storage on tilled soil surfaces. Trans. of ASAE 27:3:729-732.

Onstad, C. A., M.L. Wolfe, C. L.Larson, and D. C.Slack. 1984. Tilled soil subsidence during repeated wetting. Trans. of ASAE, 27:3:733-736.

Rawls, W. J., D. L. Brakensiek, and B. Soni. 1983. Agricultural management effects on soil water processes. Part I: Soil water retention and Green and Ampt infiltration parameters. Trans. of ASAE 26:1747-1752.

Reicosky, D. C., W.B. Voorhees, and J. K. Radke. 1981. Unsaturated water flow through a simulated wheel track. Soil Sci. Soc. Am. J. 45:3-8.

Sillanpaa, M. and L. R. Webber. 1961. The effect of freezing-thawing and wetting-drying cycles on soil aggregation. Can. J. Soil Sci. 41:182-187.

Soil Conservation Society of America. 1982. Resource Conservation glossary. Ankeny, IA.

Sommer, C. 1976. The susceptibility of agricultural soils to compaction. Grundl. Landtecknik., 26:14-23.

Taylor, J. H., E. C. Burt, and A.C. Bailey. 1980. Effect of total load on subsurface soil compaction. Trans. of ASAE 23:568-570.

Van Doren, D. M. and R. R. Allmaras. 1978. Effect of residue management practices on soil physical environment, microclimate, and plant growth. In Crop Residue Management Systems. ASA Spec. Pub. No. 31, p. 49-83.

Voorhees, W. B. 1977. Soil compaction: Our newest natural resource. Crops and Soils 29(5):13-15.

Voorhees, W. B., R. R. Allmaras, and C. E. Johnson. 1981. Alleviating tempreature stress. pp. 217-266. In G. F. Arkin and H. M. Taylor (eds.) Modifying the Root Environment to Reduce Crop Stress. ASAE Monograph, St. Joseph, MI.

Voorhees, W. B., W.W. Nelson, and G. W.Randall. 1986. Extent and persistence of subsoil compaction caused by heavy axle loads. Soil Sci. Soc. Am. J. 50:428-433.

Voorhees, W. B., C. G.Senst, and W. W. Nelson. 1978. Compaction and soil structure modification by wheel traffic in the Northern Corn Belt. Soil Sci. Soc. Am. J. 42:344-349.

Voorhees, W. B., R. A. Young, and Leon Lyes. 1979. Wheel traffic and considerations in erosion research. Trans. of ASAE 22:786-790.

Warkentin, B. P. 1971. Effects of compaction on content and transmission of water in soils. pp. 126-153. In K. K. Barnes, W. M. Carleton, T. M. Taylor, R.I. Throckmorton, and G. E. Vanden Berg (eds.) Compaction of Agricultural Soils. ASAE Monograph, St. Joseph, MI.

Wolfe, M. L. 1986. Predicting infiltration into tilled soils subject to surface sealing and consolidation using a Green-Ampt model. Unpublished Ph.D. Thesis, University of Minnesota. 190 pp.

Young, R. A. and W. B. Voorhees. 1982. Soil reosion and runoff from planting to canopy development as influenced by tractor wheel traffic. Trans. of ASAE 25:708-712.

CHAPTER 6

HYDROLOGIC EFFECTS OF CONSERVATION TILLAGE AND THEIR IMPORTANCE RELATIVE TO WATER QUALITY

J. L. Baker,
Iowa State University, Ames, Iowa

INTRODUCTION

One of the charges for this workshop was to keep in mind that we are looking at processes associated with the fate and transport of chemicals (pesticides and nitrogen) that may pose a threat to surface and groundwater resources. In this brief review, I will emphasize some particular aspects of hydrology that I believe should be clearly understood by anyone wanting to make judgements on the effects of conservation tillage on water quality.

As I view this topic, it seems that it could be simplified to one primary question: Is there less surface runoff, and/or more subsurface drainage (leaching water) with conservation tillage? And the answer is: Yes, no, well maybe. Before getting into the reasons for that type of an answer, I want to pose a secondary question and illustrate why these questions and this topic are probably more complex than first thought and are so critical to water quality concerns. As others point out, the process of water infiltration into soil is very important in determining the hydrology of a soil system; therefore a secondary question might be: How does conservation tillage affect the volume, timing, and route of infiltrating water?

The importance of the volume of infiltrating water is quite obvious in that as the volume of infiltration over the year (or growing season) increases, the volume of runoff decreases, and therefore the carrier for chemicals lost in surface runoff, where losses equal concentration times carrier, decreases.

Effects of Conservation Tillage on Groundwater Quality: Nitrates and Pesticides, Terry J. Logan et al., eds.

Conversely, as the volume of infiltrating water increases, potentially the volume of subsurface drainage, or leaching water, increases. But there is much more to it than just the volumes involved. How differences in infiltration rates caused by tillage practices change with time, both short-term (during a single storm event) and long-term (over the growing season or year), can significantly affect chemical losses. The route of percolating water, relative to chemical location, is also important.

DISCUSSION

Several studies (Baker et al., 1978; Barisas et al., 1978; and Baker et al., 1982), as well as mathematical modeling, have shown that chemical concentrations in surface runoff decline with time during a storm event. The understanding that has developed is that there is a thin "mixing zone," maybe 1 or 2 cm deep (Ahuja and Lehman, 1983), or possibly a non-uniform layer (Heathman et al., 1986), that is the main source of chemical released to water (rainfall) entering that "mixing zone," and either running off or infiltrating into deeper soil. As chemical is removed from this "mixing zone" with water, the result is lower and lower concentrations. One of the key factors in determining how quickly removal takes place is the degree of chemical soil adsorption, often defined quantitatively by the adsorption coefficient, K. For chemicals with low-to-medium K values (0 to 20), how much infiltration takes place before runoff begins is very important in determining how much chemical is lost with surface runoff.

The effect of corn residue on a disked soil surface on initial infiltration and total runoff is given in Table 1 for a rainfall simulation study. As shown, the protection provided against surface sealing by increasing amounts of corn residue resulted in the volume of infiltration occurring, before runoff began, increasing from 12 mm for the treatment with no residue to 32 mm for the highest level of residue. Overall for the 375, 750, and 1500 kg/ha corn residue treatments, initial runoff was delayed 6, 9, and 19 minutes, and total runoff voulumes were reduced to 83, 69, and 28%, respectively, relative to that for a bare disked area. As just discussed, the impact on chemical losses in runoff water was even greater than on runoff volumes (e.g., for the herbicide alachlor, losses were 57, 41, and 13% of that from the bare area). Because of initial infiltration, and contrary to popular belief, a soluble, non-adsorbed (K=0) chemical like nitrate usually experiences only minor losses with surface runoff from well-structured soils that have a sustained high infiltration rate during the initial part of a storm event.

Table 1. Effect of Residue on Disked Soil Surface on Runoff (Baker et al., 1982).

Residue kg/ha	Time [1] min.	Initial Infiltration [1]	Runoff
		mm	
0	11	12	63
375	17	18	52
750	20	21	44
1500	30	32	18

1. Time and infiltration to beginning of runoff; 127 mm rain in 2 h; plots 5% slope on sil soil.

Further evidence for the importance of residue in protecting tillage-loosend soil is found in the work of Mukhtar et al. (1985). This study involved the Paraplow, a newly introduced tillage tool in North America (Pidgeon, 1983), which lifts the soil at an angle and then drops it, which cracks and loosens the soil along natural patterns, but does not invert the soil as a moldboard plow does. In effect the soil has been loosened, but left with the residue cover of a no-till situation. Measurements of infiltration rates over the growing season showed that the increased storage and conductivity created by the soil loosening persisted for the Paraplow treatment, with infiltration rates generally higher than for the other tillage systems studied, moldboard plow, chisel plow, and no-till.

The effect of surface conditions (i.e. roughness, porosity, etc., but exlusive of residue) caused by the timing and type of tillage on initial infiltration and total runoff is shown in Table 2 for a rainfall simulation study on areas from which residue was removed. The increased roughness and porosity created by tillage increased initial infiltration and decreased total runoff relative to no-till. As is also shown, the more recent tillage (spring sweep) significantly increased initial infiltration over "aged" (fall) tillage surfaces. However, as evidenced by only a small difference in total runoff for the three tilled areas for the 2-h storm, the effect was apparently not long-lived.

Table 2. Effect of Surface Conditions (minus residue) on Runoff (Cogo et al., 1984).

Practice	Time[1] min.	Initial Infiltration[1]	Runoff
		------------ mm ------------	
Fall Plow	11	12	67
Fall Chisel-plow	13	14	64
Spring Sweep	37	39	59
No-till	3	3	102

1. Time and infiltration to beginning of runoff; 128 mm in 2 h; plots 4.5% slope on sil soil.

The effect of tillage-induced increased infiltration, but sometimes of short duration, is evident in other studies. For example, in a five-year study of runoff from adjacent row-cropped watersheds (Hamlett et al., 1984; 21 less-severe events, with one exclusion), if an event occured after all primary and secondary tillage had been performed, but before any cultivation, the watershed that had been moldboard or chisel-plowed, as opposed to the site that was only disked, had the least if any runoff. If cultivation had been performed on either or both of the watersheds, the more recently cultivated watershed had the least runoff. However, for four severe rainfall events, ranging from 42 to 74 mm, expected differences in runoff volumes from the two watersheds were evened out and runoff volumes were nearly identical.

In another study (Baker and Laflen, 1983a), rainfall simulations were performed 6 and 18 days after fall tillage on chisel-plow, disk, and no-till plots that had been grown to soybeans. For the first simulation, 2.5, 10, and 15% of the 111 mm of rain ran off for the chisel plow, disk, and no-till treatments, respectively. For the second simulation, performed on the same plots, the corresponding values for 72 mm of rain were 13, 48, and 19%. The contrast between the disk and no-till treatments was particularly striking, with runoff from the disked plots going from only 2/3 that of the no-till plots for rain one, to almost 3 times that of the no-till plots for rain two. It is believed leveling and surface sealing from rainfall impact was mainly responsible for the change.

The long-term (from storm event to storm event) changes in differences in infiltration rates are important when considering chemicals that dissipate over the growing season. As several reviews indicate (Wauchope, 1978; Baker, 1980; and Baker and Laflen, 1983b) it is usually the first storm after chemical application that results in the greatest percentage of chemical lost with surface runoff. Therefore it is conceivable that a tillage system which annually has the most runoff, may not have the largest chemical loss if for that system, chemical application closely follows a tillage practice that temporarily increases the relative infiltration rate.

In reviewing several runoff studies under simulated and natural rainfall, some patterns of the gross effects of Tillage conservation tillage on runoff volumes begin to emerge (in answer to part of our primary question; but with some of the inconsistencies noted by others). Data for the rainfall simulation studies, shown in Table 3, indicate that conservation tillage usually reduces surface runoff volumes between 10 and 50% relative to moldboard plowing. Furthermore, with only one exception, the chisel plow system resulted in a greater reduction than no-till, and no-till treatments sometimes exhibited more runoff than conventional tillage. However, rainfall simulation studies usually are short-term studies, which may mean they are run on areas with recently established tillage systems. In addition, they are usually run with long duration, intense rainfalls (e.g., 63.5 mm/h for 2 h), which as indicated earlier can even out differences that would otherwise exist for smaller rains.

Results of studies under natural rainfall are shown in Table 4. These studies, most of which involved no-till, give a slightly different picture than the rainfall simulation studies. The longer-term nature of these studies means that the tillage system had been established for a while, and under natural rainfall, rains of all descriptions can occur. As a result, no-till on the average (but again with a wide range) reduced runoff by over 50% with the very limited data possibly showing the chisel plow system to be less efficient than no-till in reducing runoff.

Use of the word efficient in describing reduction of runoff may not be appropriate because a major concern that exists is that the water that does not run off but instead infiltrates and becomes part of soil storage can percolate through the soil leaching chemicals with it. This is where the volume and route of infiltrating water become important. It is logical that the more infiltration that takes place, the more water that will potentially drain from the root zone. Although there is a deficiency of data, there are a few studies (Tyler and Thomas, 1977; Phillips, 1981; and Gold and Loudon, 1982) that have shown this to be the case for conservation tillage systems with presumeably more infiltration (and less soil-water evaporation), see Tables 5 and 6. Increased infiltration through macropores is possible, particularly with the no-till system. These macropores, resulting for example from soil cracking, root channels, or insect activity, can substantially influence the hydrology of a watershed. As an example, in one Ohio study (Harold, et al., 1970) runoff from a no-till field was only 9% that of a conventionally tilled area. Concern has been expressed that these preferential flow paths could enhance chemical movememt below the root zone. As W. M. Edwards discussed in a recent issue of U.S. Water, a no-till watershed may have "more than a thousand earthworm channels for every square foot of soil, providing a conduit for the quick flow of water borne fertilizers and pesticides below the reach of plant root systems."

Table 3. Runoff as Affected by Conservation Tillage (rainfall simulation studies).

Study[1]		Soil Texture	Slope	Runoff[2]	Note	Reference
			------ %	------		
Till-plant	(IN)	sil	8-12	86	cont. corn	Romkens et al. (1973)
Chisel				49		
Disk				85		
No-till (coulter)				74		
Till-plant	(IA)	sil	5-12	83	cont. corn	Laflen et al. (1978)
Chisel				96		
Disk				84		
No-till (Ridge)				77		
No-till (coulter)				75		
Chisel-disk	(IA)	sil	11	87	row crop	Laflen and Colvin (1981)
No-till				109		
Chisel-disk		sl	5	69		
No-till				85		
Disk-Chisel	(IL)	sil	5	72	corn	Siemens and Oschwald (1978)
Coulter-chisel				65		
Chisel				70		
Disk				70		
No-till				90		
Chisel	(WI)	sil	6	37	cont. corn	Andraski et al. (1985)
Till-plant				54		
No-till				51		
Chisel	(WI)	sil	4-6	69	cont. corn	Mueller et al. (1984)
No-till				136		
No-till	(OH)	sil	3	83	cont. corn	Van Doren et al. (1984)

1. Location of study.
2. Runoff is % of that from conventional.

Table 4. Runoff Volumes as Affected by Conservation Tillage (natural rainfall studies).

Study[1]		Soil Texture	Slope	Runoff[2]	Note	Reference
			------ %	------		
Till-plant	W(IA)	sil	10–15	65	cont. corn	Johnson et al. (1979)
No-till (ridge)				58		
No-till	W(OH)	sil	9	9	cont. corn	Harold et al. (1970)
No-till	W(OH)	sil	8–22	4	cont. corn (1972)	Triplett et al. (1978)
No-till	P(MS)	sil	5	51	beans-beans	McDowell and McGregor (1980)
No-till				38	beans-wheat	
No-till				106	beans-corn	
No-till				80	corn-beans	
Till-plant	P(SD)	sici	6	71	corn	Onstad (1972)
No-till	P	[3]	5	9	beans	Lal (1976)
No-till			15	11		
Chisel Plow	W(MI)	ci	1	80	corn-beans	Gold and Loudon (1982)
No-till	P(PA)	sici	14	7	cont. corn	Hall et al. (1984)
No-till	W(MD)	1	6–7	19	cont. corn	Angle et al. (1984)

1. W denotes watershed, P denotes plots; location in parentheses.
2. Runoff is % of that from conventional.
3. Study from Nigeria; no texture information.

To envision this difference between no-till and conventional tillage, consider two finely-porous, ceramic columns, one representing a column of soil under conventional tillage, the other, which has small, say 2 mm, diameter holes drilled through its length, representing no-till. If the columns were wetted, a nitrate salt then added to the top, and the columns flushed with ponded distilled (or low nitrate) water, the "macropores" in the no-till column would allow the nitrate to come through faster and in larger amounts. This concept was used by Tyler and Thomas (1977) to explain larger nitrate leaching losses in lysimeters planted to no-till corn than for conventional corn.

Table 5. Estimated Evaporation, Transpiration, and Subsurface Drainage (Phillips, 1981).[1]

Practice	Evap.	Trans.	Total ET	Drainage
	cm			
Conv. till	19.5	24.0	43.5	2.9
No-till	4.5	29.3	33.8	6.2

1. For cont. corn; May-Sept., 70-71; 55.0 cm rain.

Table 6. Surface and Subsurface Drainage as Affected by Tillage (Gold and Loudon, 1982).[1]

Practice	Surface Runoff	Subsurface Tile Flow
	cm	
Conventional	19.9	13.7
Chisel Plow	16.0	16.6

1. 4-ha watersheds; 0.8% slope, cl soil; 3/1/81-11/1/82.

However, now consider the case where, rather than adding the nitrate in the form of a salt at the column surfaces, the two columns are wetted with an equal amount of a solution containing nitrate, and then distilled water is ponded on the columns. In this example, the nitrate concentrations in outflow from the no-till columns would be less. In this latter example, the effect of water bypassing much of the water in the column containing nitrate was used by Wild (1972) to partially explain the slow leaching of nitrate originating from mineralization within aggregates in a fallow soil. This reasoning was also used by Kanwar et al. (1985) to explain why only 9% of the nitrate originally present was retained in the top 30 cm of conventionally tilled soil after application of 19 cm of simulated rain, while 33% remained for no-till.

SUMMARY

In summary, the answer to the question does conservation tillage decrease surface runoff and/or increase subsurface drainage is tied very closely to the effect conservation tillage has on infiltration, and is: Yes, no, well maybe. The answer is usually yes when long term comparisons are made, e.g., for all storms over a growing season or year. The answer would probably be no for small rainstorms that occur shortly after tillage, particularly in a comparison between moldboard plowing and no-till. The maybe comes in for cases where the large number of factors affecting infiltration combine in such a way to give inconsistent results, or for some large, intense storms which may "drown out" expected differences.

It needs to be emphasized that not just the volume, but the timing and route of infiltrating water is very important relative to chemical losses in surface runoff and subsurface drainage which may recharge aquifers or be intercepted by shallow drain tubes. The volume of drainage water is important in determining losses, but for surface-applied chemicals that are not strongly adsorbed, initial infiltration before runoff begins can be an important factor in decreasing concentrations and losses in surface runoff. Similarly for chemicals that degrade or dissipate with time after application, infiltration rates for the first few storms after application are important in determining surface runoff losses. On the other hand the infiltration that decreases surface runoff may enhance leaching losses. The route that leaching water takes through the soil profile (e.g., through macropores) relative to the location of a chemical of concern can also strongly influence leaching losses.

Obviously field hydrology, and changes in it caused by conservation tillage, are very important in determining the quality and quantity of agricultural drainage. Crop, soil, residue, and chemical management practices need to be manipulated in order to obtain the "best" system. However at this time all of the management options and information needed to make decisions to achieve the "best" system are not available. Compromises may have to be made, and at the moment we may have to settle for a "better" system.

REFERENCES

Andraksi, B. J., D. H. Mueller, and T. C. Daniel. 1985. Effects of tillage and rainfall simulation date on water and soil losses. Soil Sci. Soc. Am. J. 49:1512-1517.

Ahuja, L. R. and O. R. Lehman. 1983. The extent and nature of rainfall-soil interaction in the release of soluble chemicals to runoff. J. Environ. Qual. 12:34-40.

Angle, J. S., G. McClung, M. S. Mcintosh, P. M. Thomas, and D. C. Wolf. 1984. Nutrient losses in runoff from conventional and no-till corn watersheds. J. Environ. Qual. 13:431-435.

Baker, J. L. 1980. Agricultural areas as nonpoint sources of pollution. Pages 275-310 in M. R. Overcash and J. M. Davidson, Eds. Environmental Impact of Nonpoint Source Pollution, Ann Arbor Sci. Publ., Ann Arbor, MI.

Baker, J. L., J. M. Laflen. 1983a. Runoff losses of nutrients and soil from ground fall-fertilized after soybean harvest. Trans. ASAE. 26:1122-1127.

Baker, J. L., J. M. Laflen. 1983b. Water quality consequences of conservation tillage. J. Soil Water Conserv. 38:186-193.

Baker, J. L., J. M. Laflen, and R. O. Hartwig. 1982. Effects of corn residue and herbicide placement on herbicide runoff losses. Trans. ASAE 25:340-43.

Baker, J. L., J. M. Laflen, and H. P. Johnson. 1978. Effect of tillage systems on runoff losses of pesticides, a rainfall simulation study. Trans. ASAE 21:886-92.

Barias, S. G., J. L. Baker, H. P. Johnson, and J. M. Laflen. 1978. Effect of tillage systems on runoff losses of nutrients, a rainfall simulation study. Trans. ASAE. 21:893-97.

Cogo, N. P., W. C. Moldenhauer, and G. R. Foster. 1984. Soil loss reductions from conservation tillage practices. Soil Sci. Soc. Am. J. 48:368-373.

Gold, A. J. and T. L. Louden. 1982. Nutrient, sediment, and herbicide losses in tile drainage under conservation and conventional tillage. Paper No. 82-2549, ASAE, St. Joseph, MI 49085.

Hall, J. K., N. L. Harwig, and L. D. Hoffman. 1984. Cyanazine losses in runoff from no-till corn in "living" and dead mulches vs. unmulched conventional tillage. J. Environ. Qual. 13:105-110.

Hamlett, J. M., J. L. Baker, S. C. Kimes, and H. P. Johnson. 1984. Runoff and sediment transport within and from small agricultural watersheds. Trans. ASAE 27:1355-1363, 1396.

Harold, L. L., G. B. Triplett, Jr., and W. M. Edwards. 1970. No-tillage corn--characteristics of the system. Agric. Eng. 51:128-31.

Heathman, G. C., L. R. Ahuja, and J. L. Baker. 1986. Test of a non-uniform mixing model for transfer of herbicides to surface runoff. Trans. ASAE 29:450-455, 461.

Johnson, H. P., J. L. Baker, W. D. Shrader, and J. M. Laflen. 1979. Tillage system effects on sediment and nutrients in runoff from small watersheds. Trans. ASAE 22:1110-14.

Kanwar, R. S., J. L. Baker, and J. M. Laflen. 1985. Nitrate movement through the soil profile in relation to tillage system and fertilizer application method. Trans. ASAE 28:1731-1735.

Lal, R. 1976. No-tillage effects on soil properties under different crops in western Nigeria. Soil Sci. Soc. Am. J. 40:762-768.

Laflen, J. M., J. L. Baker, R. O. Hartwig, W. F. Buchele and H. P. Johnson. 1978. Soil and water loss from conservation tillage systems. Trans. ASAE 21:881-885.

Laflen, J. M., and T. S. Colvin. 1981. Effect of crop residue on soil loss from continuous row croping. Trans. ASAE. 24:605-9.

McDowell, L. L., and K. C. McGregor. 1980. Nitrogen and phosphous losses in runoff from no-till soybeans. Trans. ASAE 23:643-648.

Mueller, D. H., R. C. Wendt, and T. C. Daniel. 1984. Soil and water loss as affected by tillage and manure application. Soil Sci. Soc. Am. J. 48:896-900.

Mukhtar, S., J. L. Baker, R. Horton, and D. C. Erbach. 1985. Soil water infiltration as affected by the use of the Paraplow. Trans. ASAE 28:1811-1816.

Onstad, C. A. 1972. Soil and water losses as affected by tillage practices. Trans. ASAE 15:287-89.

Phillips, R. E. 1981. Soil moisture. In: No-tillage Research: Research Reports and Reviews, R. E. Phillips, G. W. Thomas, and R. C. Blevins, eds. University of Kentucky, College of Agriculture and Agricultural Experiment Station, Lexington.

Pidgeon, J. D. 1983. "Paraplow" a new approach to soil loosening. Paper No. 82-2136, ASAE., St. Joseph, MI 49805.

Romkens, M. J. M., D. W. Nelson, and J. V. Mannering. 1973. Nitrogen and phosphorus composition of surface runoff as affected by tillage method. J. Environ. Qual. 2:292-295.

Siemens, J. C. and W. R. Oschwald. 1978. Corn-soybean tillage systems: Erosion control, effects on crop production costs. Trans. ASAE. 21:293-302.

Triplett, G. B. Jr., B. J. Conner, and W. M. Edwards. 1978. Transport of atrazine and simazine in runoff from conventional and no-tillage corn. J. Environ. Qual. 7:77-84.

Tyler, D. D., and G. W. Thomas. 1977. Lysimeter measurements of nitrate and chloride losses from soil under conventional and no-tillage corn. J. Environ. Qual. 6:63-66.

Van Doren, D. M. Jr., W. C. Moldenhauer, and G. B. Triplett Jr. 1984. Influence of long-term tillage and crop rotation on water erosion. Soil Sci. Soc. Am J. 48:636-640.

Wauchope, R. D. 1978. The pesticide content of surface water draining from agricultural fields — a review. J. Environ. Qual. 7:459-472.

Wild, A. 1972. Nitrate leaching under bare fallow at a site in northern Nigeria. J. Soil Sci. 23:315-324.

CHAPTER 7

SOIL CHEMICAL AND BIOLOGICAL PROPERTIES AS AFFECTED BY CONSERVATION TILLAGE: ENVIRONMENTAL IMPLICATIONS

W. A. Dick,
The Ohio State University, Wooster, Ohio

T. C. Daniel,
University of Wisconsin, Madison, Wisconsin

INTRODUCTION

The application of various forms of tillage pratices (Table 1) during crop production is rapidly expanding in the United States. In 1985, 31% of the total cropland was planted using conservation tillage (CT) practices, representing a 2% increase over 1984 figures (CTIC, 1986). Projections for the year 2010 indicate approximately 95% of all cropland in the United States will be under some form of CT (Myers, 1983). The shift from conventional tillage (CnT) to CT practices has been spurred by farmers' awareness that CT is highly effective in controlling erosion and the desire to boost net profit by reducing fuel and labor costs (Ladewig and Garibay, 1983).

Until recently little information was available describing how the soil environment was changed when the moldboard plow was discarded in favor of a CT technique. It is now known that CT practices create substantial changes in physical, chemical, and biological properties within the soil profile. With application of no-tillage (NT), an increase in the concentration of nutrients at the soil surface results. This is attributed to nutrient uptake by plant roots and their subsequent deposition on the soil surface as residue and to the broadcast application of fertilizer. Pesticides may also be greatly concentrated in the soil surface layer due to lack of mixing of the soil.

Effects of Conservation Tillage on Groundwater Quality: Nitrates and Pesticides, Terry J. Logan et al., eds.

Table 1. Tillage Systems Defined[1].

Tillage System	Description
Conventional Tillage (CnT)	Moldboard plow plus at least one other secondary tillage operation prior to planting.
Conservation Tillage (CT)	Any tillage and planting system that maintains at least 30% of the soil surface covered by residue after planting.
No-Till (NT)	One of several forms of CT, usually characterized by high residue cover (60-80%) and limited field trips. The soil is left undisturbed prior to planting. Planting is completed in a narrow seedbed 2.5-7.5 cm wide. Weed control is accomplished primarily by herbicides.

1. Source of tillage system descriptions is the Executive Summary of the 1985 National Survey of Conservation Tillage Practices (CTIC, 1986).

From an environmental viewpoint, the concentration of plant nutrients and pesticides in a narrow zone of surface soil may be undesirable. Runoff of surface water which has interacted with the nutrient-enriched layer of soil will have increased concentrations of nitrates and pesticides. Initial results have indicated that surface runoff water from CT compared to CnT fields contains higher concentrations of nitrates and soluble P (Baker and Laflen, 1983; Logan and Adams, 1981; McDowell and McGregor, 1980; Romkens et al., 1973). However, total loads of P, including algal available P, may be significantly less in runoff water from CT compared to CnT fields (Andraski et al., 1985b). Reduced surface runoff and fertilizer placement were considered important parameters which produced these results (Andraski et al., 1985a; 1985c). Although pesticide losses from fields seem to be strongly related to the time between application and the first runoff event, tillage does exert a secondary effect (Baker and Johnson, 1979). Application of CT practices generally results in increased pesticide concentrations in runoff water and sediment compared to CnT practices (Baker and Johnson, 1979; Baker et al. 1978).

Chemical and biological properties of soil as affected by CT practices may also result in changes in groundwater quality. Biological activity, such as earthworm and root migration through the soil, creates macropores that, at least in the case of NT, can remain undisturbed for years after their formation (Ehlers, 1975). Water that has interacted with the nutrient- and pesticide-enriched surface soil could potentially percolate through the upper portion of the soil profile (Edwards and Norton, 1985; Ehlers, 1975) and arrive in the subsurface soil with increased concentrations of nitrates, pesticides, and other chemicals.

To better understand the effect of tillage on surface and groundwater quality, it is therefore, imperative that we accurately define the soil chemical and biological properties under the various types of tillage systems. This paper describes changes in the distribution of plant nutrients in soil profiles brought about by the application of CT practices, especially NT, and a brief overview of the effect of CT on biological properties (i.e. earthworm activity, plant diseases, and insects). The changes described provide a context in which water quality, as affected by tillage, will be discussed.

CHANGES IN SOIL CHEMICAL PROPERTIES

The redistribution or stratification of organic matter and plant nutrients within the soil profile is commonly observed to be associated with CT practices. The extent and rate to which this phenomenon develops depends on several variables including the type of CT and the length of time the tillage practice is applied, soil type, crop rotation, and fertilizer management techniques. How these variables interact with CT to effect changes in soil profile properties will be discussed in a later section of this paper. For now, a description of the soil profile that results as a direct effect of applying CT practice will be presented.

In general chemical constituents accumulate within the surface layer of the soil under CT compared to CnT (Table 2). However, below a soil depth of approximately 5 to 10 cm the pattern is often reversed. Stratification of organic matter, especially under NT, has been observed repeatedly (Blevins et al., 1977; 1983; Dick, 1983; Doran, 1980; Van Doren et al., 1977; Juo and Lal, 1979; Shuman and Hargrove, 1985; Groffman, 1985). A two-fold reason for this observation can be put forward: 1) under CT the majority of the crop residues are maintained at the soil surface instead of being mixed throughout the tillage layer; and 2) plowing and secondary tillage increase the rate of organic matter loss because tillage stimulates greater microbial contact with the residue and thus greater microbial activity. However, the increased microbial activity under CnT lasts only temporarily until the readily available organic residue has been consumed to produce CO_2 or stabilized humic compounds.

Table 2. Organic C and N[1] and double-acid extractable nutrients[2] (kg/ha) in conventional-tillage (CnT) and no-tillage (NT) soils at Horseshoe Bend.

0-5	9067 *	14155	849 *	1201	21 *	37	103 *	176	314 *	709	50 *	92
5-13	11481	10672	1306	1234	28	15	167	158	580	505	94	75
13-21	741	4942	962	815	12	6	121	106	411	283	66	46
0-21	27689	29769	3117	3250	61	58	391	440	1305	1497	210	213

1. Annual means from 1983.
2. Data from June 1983.
3. Hendrix et al. (1986).
4. Asterick (*) indicates tillage treatments differ significantly at P = 0.05.

The extent of stratification brought about by a change in tillage depends on many soil variables but it is not uncommon to observe a doubling of the organic matter content in the surface few centimeters of soil under NT compared to CnT (Fig. 1). The distribution of other organic constituents in NT soil, such as organic N and P, parallels closely that observed for organic C (Dick, 1983).

Acidity produced by nitrification of ammonium ions derived from fertilizer or organic matter is generally greater at the soil surface where CT compared to CnT practices are maintained (Fig. 2). However, immediately after a broadcast application of lime, the surface pH will be higher under CT than under CnT. When CT practices bring about a combination of low pH values and high organic matter concentrations, the effectiveness of certain herbicides may be reduced. If this loss of efficacy is compensated by increasing rates of application to ensure adequate weed control, it becomes readily apparent how tillage practices may ultimately affect other factors, e.g. surface water quality.

Probably the plant nutrient most affected by change in tillage is P as it is essentially immobile in soil. Concentration of plant available P in the surface layer (0-1.25 cm) of soil no-tilled continuously for 18 years, and to which the majority of fertilizer was broadcast applied, was 8 times higher than in a similarly fertilized CnT soil (Fig. 3). Although this represents an extreme case, other studies have shown increased available P levels resulting from CT. Even when CT practices have been applied for 6 years or less, it is not uncommon to see soil test levels of available P two to three times greater under CT as compared to when CnT is practiced (Eckert and Johnson, 1985; Moschler et al., 1972; Triplett and Van Doren 1969; Fink and Wesley, 1974). Increase in available P concentration at the soil surface caused by CT fertilizer

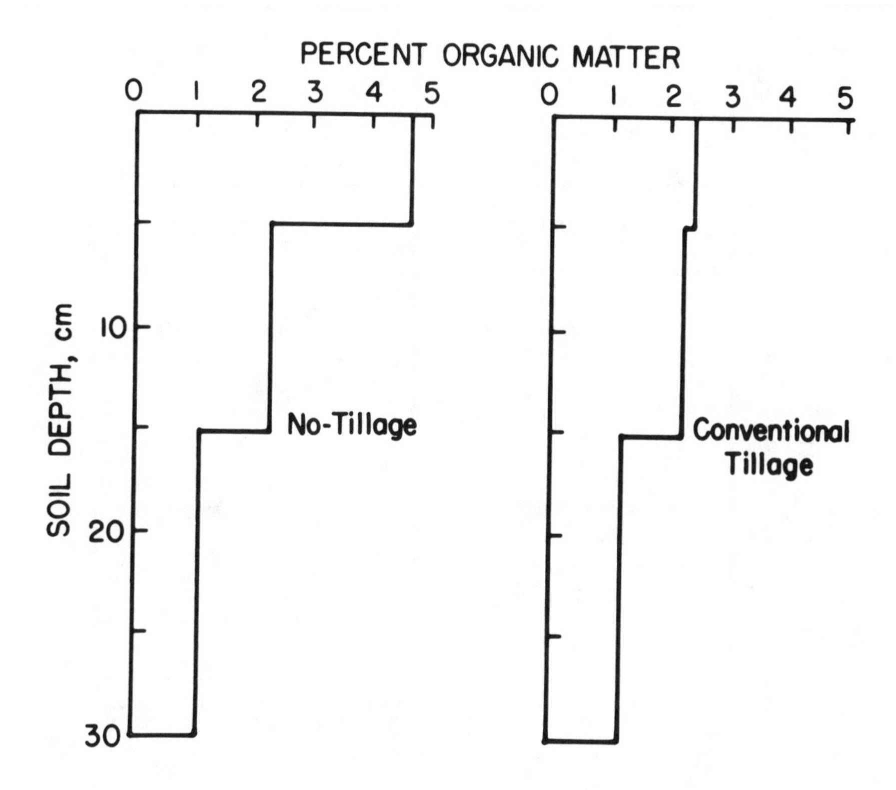

Figure 1. Soil organic matter distribution after 10 years of no-till and conventional tillage corn production (from Blevins et al., 1983).

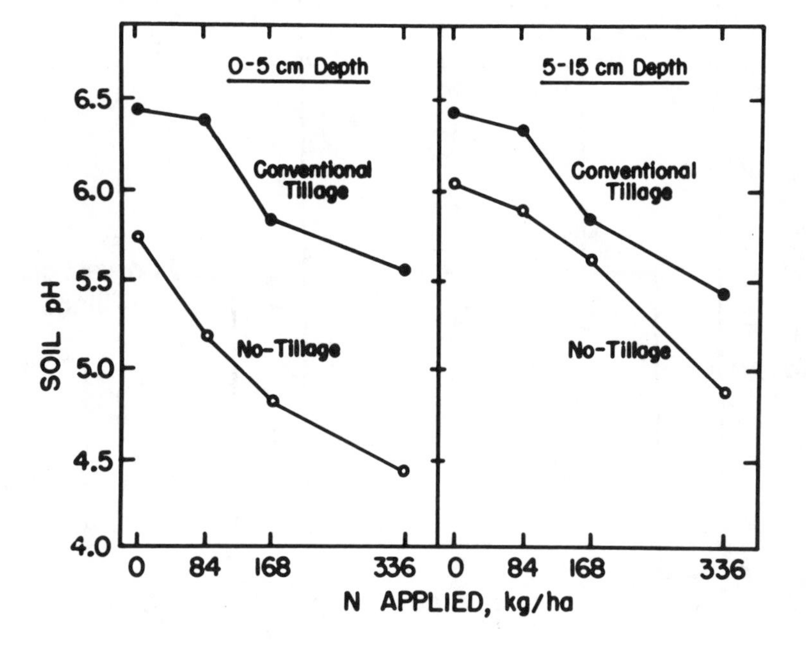

Figure 2. Soil pH at two soil depths after 10 years of corn production by no-till and conventional tillage under different levels of N fertilizer (from Blevins et al., 1983).

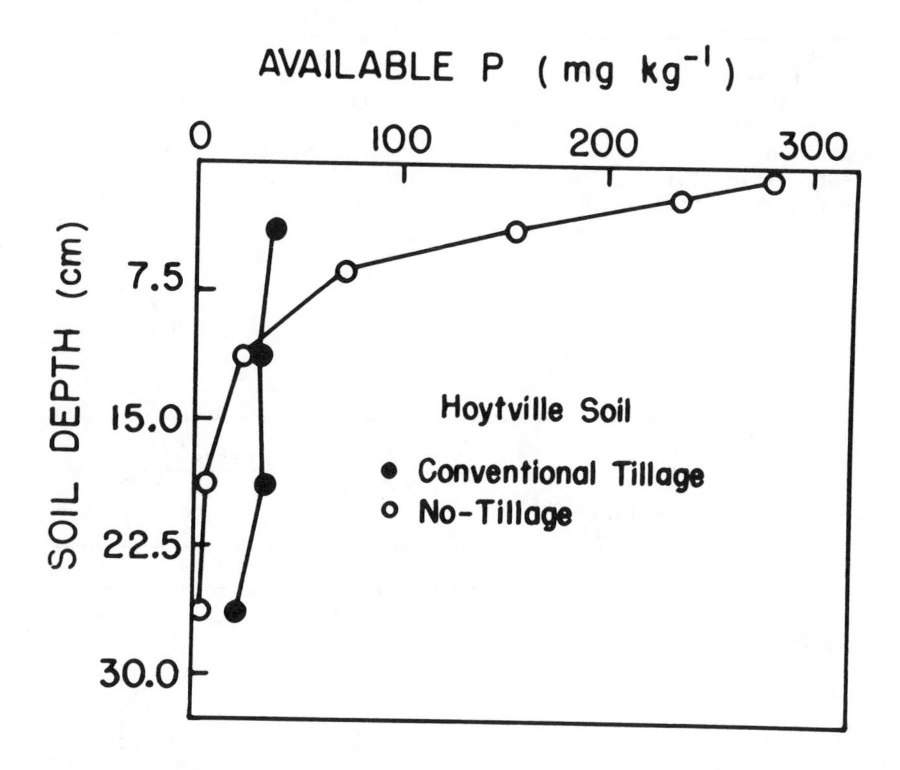

Figure 3. Available P concentrations in a soil profile as affected by tillage (from Dick et al., 1986a).

management practices provides an explanation for the reported increased soluble P losses under CT (Baker and Laflen, 1983; Logan and Adams, 1981; McDowell and McGregor, 1980; Romkens et al., 1973). Water flowing over the surface of CT fields can potentially interact with the P-enriched layer, bringing higher concentrations of P into solution than from CnT fields. However, as Andraski et al. (1985a; 1985b) demonstrated, fertilizer management plays a key role in determining whether P loadings are increased under CT regardless of the existence and extent of stratification. While dissolved molybdate-reactive P loadings were unaffected by tillage, the more appropriate water quality parameter of algal available P loadings was significantly reduced under CT conditions (Table 3). Fertilizer applied in the row of planting as opposed to broadcast application resulted in a symbiotic relationship between a sound production practice and preservation of surface water quality.

Table 3. Effect of tillage on soil test levels and P loading contained in surface runoff.[1]

Tillage	Available P[2]	Loading rates: Dissolved molybdate-reactive P	Loading rates: Algal available P
	mg/kg	mg/m^2	mg/m^2
Conventional	39	2.9[a3]	17.8[a]
Chisel	48	1.5[a]	10.3[bc]
No-till	62	2.3[a]	6.6[c]

1. Andraski et al. (1985a; 1985b).
2. Bray P-1 concentrations in the 0-2.5 cm soil layer.
3. Values followed by the same letter are not significantly different at P = 0.1.

Stratification of other nutrients such as K, Ca, Mg, Mn, Fe, Cu, and Zn has also been associated with the adaptation of CT (Hendrix et al., 1986; Shuman and Hargrove, 1985; Blevins et al., 1983; Chandi and Takkar, 1982; Hargrove et al., 1982; Lal, 1976). The solubility and availability of Ca, Mg, and the micronutrients are strongly influenced by pH so that the application of soil-acidifying fertilizers or lime plays a major role in the soil solution concentrations observed. As a result, it is difficult to generalize as to whether concentrations will be higher under CT or under CnT. The stratification of various elements in soil may not always have a direct effect on water quality but may have an indirect effect by promoting or inhibiting biological activity or altering the chemistry of the soil so that the interaction of fertilizers and pesticides with the soil is changed.

An important element from both an agricultural and environmental viewpoint is N. As previously mentioned, organic N is stratified in soils to which CT practices have been applied. Mineral N concentrations in soil as affected by tillage have also been measured. Dowdell et al. (1983) found nitrate concentrations in the soil solution were greater in plowed fields than in direct-drilled plots. This difference occurred during the months immediately following fall plowing and by spring no significant differences could be detected. Results similar to those of Dowdell have been reported by others (Rice and Smith, 1983; Linn and Doran, 1984; Carter and Rennie, 1984).

It is perhaps not surprising that soil disturbance (e.g. plowing) causes increased nitrate concentrations in the soil, presumably due to increased organic N mineralization. Yet despite the fact that nitrate concentrations may be similar or even lower under CT compared to CnT, concentrations of nitrate in surface runoff water from CT fields have often been observed to greater (McDowell and McGregor, 1980; Baker and Laflen, 1983). The interaction between tillage and soil nitrate concentrations has been studied by Rice and Smith (1983). They determined that production of nitrate via nitrification of ammonium fertilizers was more rapid in NT soils. Romkens et al. (1973) have indeed shown that runoff water contained higher levels of nitrate due to surface application of fertilizer. However where no ammonium fertilizer additions are made, ammonium concentrations become more limiting under CT compared to CnT systems (Rice and Smith, 1983). This results in lower nitrate concentrations under CT and presumably lower nitrate concentrations in surface water runoff. In addition, tillage also stimulates organic nitrogen mineralization increasing ammonium concentrations and ultimately nitrate concentrations.

CHANGES IN SOIL BIOLOGICAL PROPERTIES

The accumulation of organic matter and inorganic nutrients in the surface soil layer under CT creates an environment where biological activity may be greatly stimulated. Hendrix et al. (1986) have hypothesized that under CnT, distribution of plant residue throughout the plowed layer promotes bacterial activity and hence abundance and activity of bacteriovorous fauna. Decomposition and nutrient mineralization proceed rapidly. Under NT, plant residue is primarily localized on the soil surface and fungal growth is promoted. The abundance and activity of fungivorous fauna increase in the surface layers and mineralization proceeds more slowly. The numbers and relative biomass of nematodes, which are among the most abundant soil fauna, support this hypothesis. Results from their study conducted in Georgia (Hendrix et al. 1986) revealed significantly greater numbers of bacteriovorous and significant lower numbers of fungivorous nematodes under CnT compared to NT systems. The numbers of herbivorous nematodes were similar under the two tillage treatments.

The effect of CT on other soil biological properties, such as soil enzyme, insect, disease and earthworm activities has also been studied. Soil enzyme activities have often been used as indices of microbial activity (Casida, 1977; Hersman and Temple, 1979; Frankenberger and Dick, 1983). In addition, soil enzymes may play an important role in the cycling of nutrients and pesticides. For example, the breakdown of the N-containing urea fertilizer is via the activity of soil urease. Activities of several enzymes in two soils in Ohio to which CnT and NT practices have been continuously applied for 18 or 19 years were much greater under NT in the surface soil layer (0-1.25 cm) than under CnT (Table 4). Generally, in the lower portion of the profile (22.5-30.0 cm soil layer), the activities of soil enzymes were greater under CnT than NT. The enzyme activities were found to be closely correlated to organic C content in the soil profile (Dick, 1984).

Studies of the effect of CT practices on insects have focused primarily on the detrimental activity of pests in corn (House and Stinner, 1983). Initially, as entomological studies on CT systems were first begun, it was thought that CT practices increased insect pest activity (Musick, 1973; Gregory and Musick, 1976). However, recent research has indicated CT practices also increase beneficial insect populations (Stinner et al., 1986; House and Stinner, 1983; House and All, 1981; All et al., 1979). The relationship between insect populations and tillage have been summarized by Gregory and Musick (1976) as follows: (i) the level of insect activity is related to the previous crop grown and (ii) NT systems more often support the simultaneous attack of insects than do CnT systems. Recent work has indicated that some pest problems may really be an indirect response to CT practices (Stinner et al., 1984). For example, stalk borer (Papaipema nebris (Gn.)) problems in NT corn were not caused directly by tillage but rather indirectly via increased orchardgrass (Dactylis glomerata) densities under NT (Fig. 4). Improved management of CT systems may reduce many of the insect pest problems currently associated with CT practices.

Because of the change in the soil-plant environment associated with CT compared to CnT, disease problems of various crops also change. Several plant pathogenic fungi and bacteria overwinter in and on residue from diseased plants of the previous years so that the residue serves as the primary source of inoculant for disease development early in the growing season (Cook et al., 1978). As a result, a new spectrum of corn diseases has emerged with CT while some of the traditional disease problems, previously associated with CnT, have decreased in severity (Martenson, 1981). Several reviews have been written in the past few years (Sumner et al., 1981; Cook et al., 1978) on the effect of reduced tillage and increased plant residue cover on plant diseases. The general consensus seems to be that CT will increase the incidence of some diseases while decreasing others, or will have no effect on plant disease.

Table 4. Effect of tillage on enzyme activities in soil profiles (from Dick et al., 1986a, 1986b).

			Soil Type	
Enzyme[1]	Soil layer (cm)	Tillage[2]	Typic Fragiudalf	Mollic Ochraqualf
Alkaline Phosphatase	0-1.25	NT	207	368
		CnT	83	203
	22.5-30.0	NT	24	198
		CnT	48	149
Amidase	0-1.25	NT	26.2	29.0
		CnT	7.5	16.4
	22.5-30.0	NT	2.7	17.6
		CnT	3.1	13.1
Invertase	0-1.25	NT	378	325
		CnT	92	63
	22.5-30.0	NT	20	20
		CnT	19	33
Urease	0-1.25	NT	460	532
		CnT	89	177
	22.5-30.0	NT	19	55
		CnT	23	126

1. Alkaline phosphatase activity is expressed as ug of p-nitrophenol released/g soil/h, amidase activity as ug of NH_3-N released/3g soil/24h, invertase activity as ug glucose released/g soil/h, and urease activity as ug NH_3-N released/g soil/4h.
2. NT, no-tillage and CnT, conventional tillage.

When evaluating the environmental implication of changes in soil biological activity caused by CT, the most important variable may be the perception of the producer using the CT practice. As CT practices developed, especially NT, increased insect, disease and other crop pests (e.g. weeds) were evident. In addition, fertilizer efficiency was also considered to be lower where NT was practiced (Meisinger et al., 1985; Thomas et al., 1973; Legg et al., 1979; Bandel et al., 1975). As a result, pesticide and fertilizer recommendations and additions were often higher for CT systems than for CnT systems. This increased loading rate may directly effect the quality of both surface and subsurface waters coming from reduced tillage fields. However, with increased research efforts and experience in managing CT systems, the ability to grow crops using CT practices without increasing inputs of fertilizers and pesticides is now available. What is needed is a reversal of the perception that the increased biological activity associated with CT is a negative factor that leads to increased fertilizer immobilization or loss, increased insect, and increased disease problems which require greater inputs of fertilizers and pesticides.

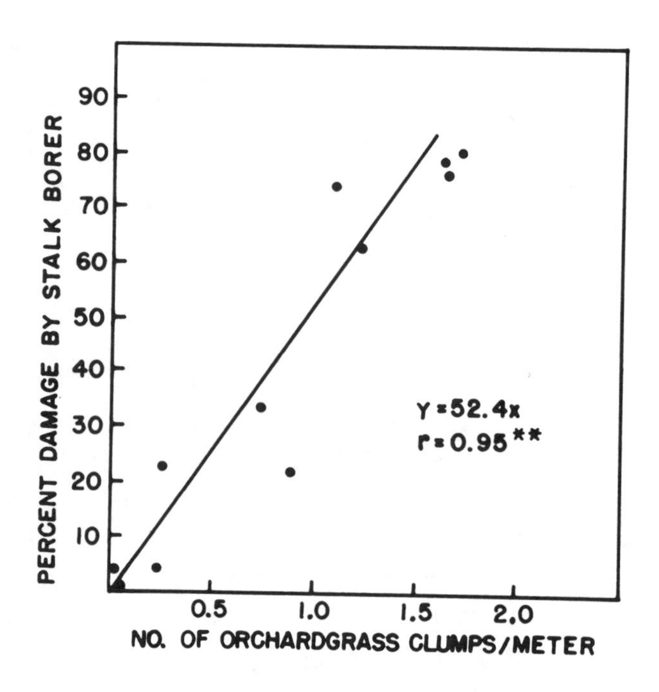

Figure 4. Relationship between stalk borer damage to corn and the number of herbicide-killed orchard grass clumps per meter of row (from Stinner et al., 1984).

One other biological parameter that is affected by CT practices and which has water quality implications is earthworm activity. Earthworm activity is important for increasing plant residue decomposition (Mackay and Kladivko, 1985), aggregate stability (Mackay and Kladivko, 1985; Hopp and Hopkins, 1946), water-holding capacity (Stockdill, 1982), pore size and infiltration rate (Ehlers, 1975), and in reducing soil crusting (Kladivko et al., 1986). Several studies have been conducted investigating the effect of reduced tillage on earthworm populations (Mackay and Kladivko, 1985; De St. Remy and Daynard, 1982; Edwards and Lofty, 1982; Barnes and Ellis, 1979). Generally the numbers of earthworms under CT are similar or only slightly increased compared to CnT (Tables 5 and 6). However, earthworm channels formed under NT are not destroyed each year by tillage and provide large channels for water infiltration. A continuous network of earthworm channels which connects the surface with the subsoil (to a depth of 180 cm) was observed by Ehlers (1975) under NT. The maximum infiltrability of these channels in the NT soil was determined to be more than 1 mm (1 liter per m^2) per minute, even though the volume of these channels amounted to only 0.2% of the total soil volume. Similar results were observed by Edwards and Norton (1985) for a soil which had been under NT corn for 20 consecutive years. Dye studies showed that even at low rainfall rates, water moved rapidly through vertically continuous worm channels.

The combination of increased concentrations of nutrients and pesticides in the surface layer of NT soil interacting with the surface water and the increased rate of transport of the water through earthworm channels into the soil subsurface could have a significant impact on groundwater quality. Since most of the biological activity is in the rooting portion of the soil and decreases rapidly with depth, movement of materials such as pesticides through earthworm channels could result in the bypassing of the biologically active rooting zone so that reduced breakdown of the pesticides occurs. Chemical sorption-desorption reactions will still occur in the subsoil but continual flushing of water through the soil may eventually cause the pesticide to move into the groundwater.

Table 5. Effect of crop and tillage practice on earthworm numbers (Mackay and Kladivko, 1985).

Crop	Tillage Practice	Total Earthworm Numbers/m^2
Corn	No-till	4.4 ± 2.0[1]
	Plow	3.1 ± 3.0
Soybean	No-till	26.5 ± 8.3
	Plow	11.9 ± 5.1
Clover/bluegrass	Borderstrip	54.9 ± 8.8
Clover/ryegrass	Pasture	224.5 ± 30

1. Standard deviations.

Table 6. Mean number of earthworms in monoculture corn as affected by tillage (De St. Remy and Daynard, 1982).

Treatment	Sampling date			
	May 9	June 5	Sept 8	Oct 6
	Number/m^2			
(Mature worms)				
Fall plow	11.2 a[1]	8.0 a	8.0 b	11.2 a
Spring plow	8.0 a	4.8 a	11.2 ab	24.0 a
No-till	14.4 a	11.2 a	24.0 a	14.4 a
(Immature worms)				
Fall plow	56.0 a	28.8 a	38.4 a	56.0 b
Spring plow	33.6 a	11.2 b	33.6 a	126.4 a
No-till	43.2 a	43.2 a	62.4 a	68.8 b

1. Means followed by the same letter within a treatment column are not significantly different at P = 0.05 (Duncan's multiple range test).

INTERACTION OF TILLAGE AND OTHER PARAMETERS ON SOIL CHEMICAL AND BIOCHEMICAL PROPERTIES

Although the type of tillage may be the dominant factor in creating changes in soil chemical and biological properties, other management and climatic factors also interact with tillage to accentuate or lessen the tillage affects (Smith and Blevins, see Chapter 8). For example, the length of time a CT practice is continuously applied affects the degree of nutrient stratification in a soil. Concentrations of organic C in soils to which NT practices have been continuously applied for five years or less indicate a rapid increase occurs resulting in concentrations almost double compared to where CnT practices have been maintained (Rice and Smith, 1982; Groffman, 1985; Doran, 1980). However, after the first years of NT, changes in soil organic matter occur more slowly. After approximately 20 years of NT application to two soils in Ohio, concentrations of organic C at the soil surface were 2.5 and 2.0 times greater than under CnT (Dick, 1983). Plant available P concentrations in soil are especially sensitive to changes in tillage. Eckert and Johnson (1985) found that significant changes in available P levels occurred in the 0-5 cm soil layer after only three years of NT, even when no P fertilizer was applied. After approximately 20 years of NT, available P levels were 7.5 and 3.5 times higher in the surface (0-1.25 cm) soil layer compared to concentrations in soil under CnT (Dick et al., 1986a; 1986b).

Soil type is also important but may be less noticeable than length of time of tillage application. Long-term tillage experiments in Ohio have shown that the comparative effects of NT and CnT extend deeper into a Mollic Ochraqualf soil than a Typic Fragiudalf soil (Dick, 1983). This was attributed to the shrink-swell properties of the Mollic Ochraqualf soil which created cracking in the soil when dry and allowed for the oxidation of organic matter to occur deeper in the soil profile. The organic matter that was oxidized in the NT soil could not be replaced by the incorporation of plant residues by tillage. The cracking may have also allowed for deeper penetration of roots resulting in greater removal of plant nutrients which were not replaced by broadcast application of fertilizer.

A specific example of how tillage and soil type may interact to affect surface water quality is illustrated by the data in Table 7. The Mollic Ochraqualf soil contains approximately 30% more total P than does the Typic Fraguidalf soil yet exhibits a lower P availability in the CnT soil. This is because of the greater P fixation tendencies of the Mollic Ochraqualf soil. As a result of this fixation, P recommendations and fertilizer rates required to achieve high crop yields are greater. However, continued broadcast application of the higher rates of P fertilizer on the soil surface have saturated the P fixation sites resulting in available P levels in the NT soil being much greater in the Mollic Ochraqualf soil compared to the Typic Fragiudalf soil.

Because of the high levels of soil surface available P, concentrations of soluble P in surface runoff from the Mollic Ochraqualf soil could be expected to be greater than from the Typic Fraguidalf soil.

Table 7. Interaction of soil type and tillage on total and available phosphorous concentrations (Dick et al., 1986a; 1986b).

Soil	Tillage[1]	Total P[2]	Available P[3]
		(mg/kg)	
Typic Fragiudalf (Wooster soil)	CnT	580	45
	NT	609	160
Mollic Ochraqualf (Hoytville soil)	CnT	867	38
	NT	868	282

1. Cnt, conventional tillage and NT, no-tillage. Tillage treatments had been continuously applied for 19 and 18 years in the Wooster and Hoytville soils, respectively.
2. Total P concentration in the 0-30 cm soil layer.
3. Available P concentration in the 0-7.50 cm soil layer (CnT) or the 0-1.25 cm soil layer (NT).

The method of fertilizer application has already been implicated (Andraski et al., 1985a; 1985b) as an important variable that may interact with tillage to affect soil chemical and biological properties. Prior to some major equipment changes the standard method of fertilizer application was broadcast. Broadcast fertilizer applications are still widely used because of their convenience. It is not surprising that repeated broadcast applications of fertilizers such as ammonium based N fertilizers or P fertilizers will create a highly stratified pH and nutrient profile. Fox and Hoffman (1981) and Blevins et al. (1983) have shown that the higher the rate of N application, the more dramatic the impact on soil pH (see Fig. 2). The pH decrease in the surface layer of a NT soil is directly proportional to the application rate of N, with

$(NH_4)_2SO_4$ fertilizer having a greater affect than NH_4NO_3, urea, or urea-ammonium nitrate (UAN) fertilizers (Fox and Hoffman, 1981).

Crop rotation interacts with tillage to affect soil chemical and biological properties because of the different management variables associated with each crop such as fertilizer and pesticide requirements and timing of fertilizer applications. Crop rotation also affects several soil properties as a result of the amount and kind of plant residues produced. The organic matter content and nitrogen mineralization potential of a Palouse silt loam (Pachic Ultic Haploxerolls) of Washington state were highest under a combination treatment of NT and a high residue-yielding rotation (alternate spring wheat and winter wheat after a green manure crop (El-Harris et al., 1983)). The lowest levels were under moldboard plow and a winter wheat-pea rotation. In Ohio, soil enzyme activities were also affected by both tillage and rotation (Dick, 1984). The highest enzyme activities were associated in NT plots with the rotation (corn-oats-meadow) which produced the greatest amount of residue and the lowest activities were associated with the rotation producing the least amount of residue (corn-soybean). In general, the effect of crop rotation on soil chemical and biological properties is more evident when coupled with NT than with plowed treatments.

SUMMARY

Conservation practices have a major impact on the chemical and biological properties of soil. When compared to CnT, the surface soil layer of CT fields, especially NT, becomes enriched in hydrogen ions, plant nutrients, and biological activity. In the lower portion of the rooting zone, from approximately 5 to 10 cm depth, the opposite trend often occurs and the NT soil has lower concentrations of the chemical and biological parameters. Tillage operations, in which the soil is inverted and mixed, tends to counteract changes in chemical and biological properties. However, with CT and most noticeably NT, the growing of annual crops and additions of fertilizer produce changes in the soil profile that are noticeable in only a few years. The effect of these changes in soil chemical and biological properties must be clearly defined and understood if we are to fully comprehend the impacts of tillage on our environment.

REFERENCES

All, J. N., R. N. Gallaher, and M. D. Jellum. 1979. Influence of planting date, preplanting weed control, irrigation, and conservation tillage practices on efficacy of planting time insecticide applications for control of lesser cornstalk borer in field corn. J. Econ. Entomol. 72:265-268.

Andraski, B. J., D. H. Mueller, and T. C. Daniel. 1985a. Effects of tillage and rainfall simulation date on water and soil losses. Soil Sci. Soc. Am. J. 49:1512-1517.

Andraski, B. J., D. H. Mueller, and T. C. Daniel. 1985b. Phosphorus losses in runoff as affected by tillage. Soil Sci. Soc. Am. J. 49:1523-1527.

Andraski, B. J., T. C. Daniel, B. Lowery, and D. H. Mueller. 1985c. Runoff results from natural and simulated rainfall for four tillage systems. Trans. ASAE 28:1219-1225.

Baker, J. L. and H. P. Johnson. 1979. The effect of tillage systems on pesticides in runoff from small watersheds. Trans. ASAE 22:554-559.

Baker, J. L. and J. M. Laflen. 1983. Water quality consequences of conservation tillage. J. Soil Water Conserv. 38:186-193.

Baker, J. L., J. M. Laflen, and H. P. Johnson. 1978. Effect of tillage systems on runoff losses of pesticides, a rainfall simulation study. Trans. ASAE 21:886-892.

Bandel, V. A., S. Dzienia, G. Stanford, and J. O. Legg. 1975. Nitrogen behavior under no-till vs conventional corn culture. I. First-year results using unlabeled nitrogen fertilizer. Agron. J. 67:782-786.

Barnes, B. T. and F. B. Ellis. 1979. Effects of different methods of cultivation and direct drilling, and disposal of straw residues on populations of earthworms. J. Soil Sci. 30:669-679.

Blevins, R. L., G. W. Thoms, and D. L. Cornelius. 1977. Influence of no-tillage and nitrogen fertilization on certain soil properties after 5 years of continuous corn. Agron. J. 69:383-386.

Blevins, R. L., M. S. Smith, G. W. Thomas, and W. W. Frye. 1983. Influence of conservation tillage on soil properties. J. Soil Water Conserv. 38:301-305.

Carter, M. R. and D. A. Rennie. 1984. Nitrogen transformations under zero and shallow tillage. Soil Sci. Soc. Am. J. 48:1077-1081.

Casida, L. E. Jr. 1977. Microbial metabolic activity in soil as measured by dehydrogenase determinations. Appl. Environ. Microbiol 34:630-636.

Chandhi, K. S. and P. N. Takkar. 1982. Effects of agricultural cropping systems on micronutrients transformation. I. Zinc. Plant Soil 69:423-436.

Conservation Tillage Information Center (CTIC). 1986. Executive Summary of the 1985 National Survey of Conservation Tillage Practices. CTIC, Fort Wayne, Indiana.

Cook, R. J., M. G. Boosalis, and B. Doupnik. 1978. Influence of crop residues on plant diseases. pp. 147-163. In Crop Residue Management Systems. Special Publication 31, American Society of Agronomy, Madison, Wisconsin.

De St. Remy, E. A. and T. B. Daynard. 1982. Effects of tillage methods on earthworm populations in monoculture corn. Can. J. Soil Sci. 62:699-703.

Dick, W. A. 1983. Organic carbon, nitrogen, and phosphorus concentrations and pH in soil profiles as affected by tillage intensity. Soil Sci. Soc. Am. J. 47:102-107.

Dick, W. A. 1984. Influence of long-term tillage and crop rotation combinations on soil enzyme activities. Soil Sci. Soc. Am. J. 48:569-574.

Dick, W. A., D. M. Van Doren, Jr., G. B. Triplett, Jr., and J. E. Henry. 1986a. Influence of long-term tillage and rotation combinations on crop yields and selected soil parameters. I. Results obtained for a Mollic Ochraqualf soil. Research Bulletin No. 1180 of The Ohio State University/The Ohio Agricultural Research and Development Center, Wooster, Ohio.

Dick, W. A., D. M. Van Doren, Jr., G. B. Triplett, Jr. and J. E. Henry. 1986b. Influence of long-term tillage and rotation combinations on crop yields and selected soil parameters. II. Results obtained for a Typic Fragiudalf soil. Research Bulletin No. 1181 of The Ohio State University/The Ohio Agricultural Research and Development Center, Wooster, Ohio.

Doran, J. W. 1980. Soil microbial and biochemical changes associated with reduced tillage. Soil Sci. Soc. Am. J. 44:765-771.

Dowdell, R. J., R. Crees, and R. Q. Cannell. 1983. A field study of effects of contrasting methods of cultivation on soil nitrate content during autumn, winter and spring. J. Soil Sci. 34:367-379.

Eckert, D. J. and J. W. Johnson. 1985. Phosphorus fertilization in no-tillage corn production. Agron. J. 77:789-792.

Edwards, C. A. and J. R. Lofty. 1982. The effect of direct drilling and minimal cultivation on earthworm populations. J. Applied Ecology. 19:723-734.

Edwards, W. M. and L. D. Norton. 1985. Characterizing macropores after 20 years of continuous no-tillage corn. Agronomy Abstracts. p. 205. American Society of Agronomy, Madison, Wisconsin.

Ehlers, W. 1975. Observations on earthworm channels and infiltration on tilled and untilled loess soil. Soil Sci. 119:242-249.

El-Harris, M. K., V. L. Cochran, L. F. Elliot, and D. F. Bezdicek. 1983. Effect of tillage, cropping, and fertilizer management on soil nitrogen mineralization potential. Soil Sci. Soc. Am. J. 47:1151-1161.

Fink, R. J. and D. Wesley. 1974. Corn yield as affected by fertilization and tillage system. Agron. J. 66:70-71.

Fox, R. H. and L. D. Hoffman. 1981. The effect of N fertilizer source on grain yield, N uptake, soil pH and lime requirement in no-till corn. Agron. J. 93:891-895.

Frankenberger, W. T. Jr. and W. A. Dick. 1983. Relationships between enzyme activities and microbial growth and activity indices in soil. Soil Sci. Soc. Am. J. 47:945-951.

Gregory, W. W. and G. J. Musick. 1976. Insect management in reduced tillage systems. Bull. Entomol. Soc. Am. 22:302-334.

Hargrove, W. L., J. T. Reid, J. T. Touchton, and R. N. Gallaher. 1982. Influence of tillage practices on the fertility status of an acid soil double-cropped to wheat and soybeans. Agron. J. 74:684-687.

Hendrix, P. F., R. W. Parmalee, D. A. Crossby, Jr. D. C. Coleman, E. P. Odom, and P. M. Groffman. 1986. Detritus food webs in conventional and no-tillage agroecosystems. BioScience 36:374-380.

Hersman, L. E. and K. L. Temple. 1979. Comparison of ATP, phosphatase, pectinolyase, and respiration as indicators of microbial activity in reclaimed coal strip mine spoils. Soil Sci. 129:70-73.

Hopp, H. and H. T. Hopkins. 1946. Earthworms as a factor in the formation of water-stable aggregates. J. Soil Water Conserv. 1:11-13.

House, G. J. and J. N. All. 1981. Carabid beetles in soybean agroecosystems. Environ. Entomol. 10:194-196.

House, G. J. and B. R. Stinner. 1983. Arthropods in no-tillage soybean agroecosystems: Community composition and ecosystem interactions. Environ. Manag. 7:23-28.

Juo, A. S. R. and R. Lal. 1979. Nutrient profile in a tropical Alfisol under conventional and no-till systems. Soil Sci. 127:168-173.

Kladivko, E. J., A. D. Mackay, and J. M. Bradford. 1986. Earthworms as a factor in the reduction of soil crusting. Soil Sci. Soc. Am. J. 50:191-196.

Ladewig, H. and R. Garibay. 1983. Reasons why Ohio farmers decide for or against conservation tillage. J. Soil Water Conserv. 38:487-488.

Lal, R. 1976. No-tillage effects on soil properties under different crops in western Nigeria. Soil Sci. Soc. Am. J. 40:762-768.

Legg, J. O., G. Stanford, and O. L. Bennett. 1979. Utilization of labeled-N fertilizer by silage corn under conventional and no-till culture. Agron. J. 71:1009-1015.

Linn, D. M. and J. W. Doran. 1984. Effect of water-filled pore space on carbon dioxide and nitrous oxide production in tilled and non-tilled soils. Soil Sci. Soc. Am. J. 48:1267-1272.

Logan, T. J. and J. R. Adams. 1981. The effects of reduced tillage on phosphate transport from agricultural lands. Lake Erie Wastewater Management Study Tech. Rep. Series. U.S. Army Engineer District, Buffalo, NY.

Mackay, A. D. and E. J. Kladivko. 1985. Earthworms and rate of breakdown of soybean and maize residues in soil. Soil Biol. Biochem. 17:851-857.

Martenson, C. A. 1981. Corn diseases and cultural practices. Iowa State University Cooperative Extension Service, Ames, Iowa.

McDowell, L. L. and K. C. McGregor. 1980. Nitrogen and phosphorus losses in runoff from no-tillage soybeans. Trans. ASAE 23:643-648.

Meisinger, J. J., V. A. Bandel, G. Stanford, and J. O. Legg. 1985. Nitrogen utilization of corn under minimal tillage and moldboard plow tillage. I. Four-year results using labeled N fertilizer on an Atlantic Coastal Plain soil. Agron. J. 77:602-611.

Moschler, W. W., G. M. Shear, D. C. Martins, G. D. Jones, and R. R. Wilmouth. 1972. Comparative yield and fertilizer efficiency of no-tillage and conventionally tilled corn. Agron. J. 64:229-231.

Musick. G. J. 1973. Control of armyworm in no-tillage corn. Ohio Report 58:42-45.

Myers, P. C. 1983. Why conservation tillage? J. Soil Water Conserv. 38:136.

Rice, C. W. and M. S. Smith. 1982. Denitrification in no-till and plowed soils. Soil Sci. Soc. Am. J. 46:1168-1173.

Rice, C. W. and M. S. Smith. 1983. Nitrification of fertilizer and mineralized ammonium in no-till and plowed soil. Soil Sci. Soc. Am. J. 47:1125-1129.

Romkens, M. J. M., D. W. Nelson, and J. V. Mannering. 1973. Nitrogen and phosphorus composition of surface runoff as affected by tillage method. J. Environ. Qual. 2:292-295.

Shuman, L. M. and W. L. Hargrove. 1985. Effect of tillage on the distribution of manganese, copper, iron, and zinc in soil fractions. Soil Sci. Soc. Am. J. 49:1117-1121.

Smith, M. S. and R. L. Blevins. 1987. Effect of conservation tillage on biological and chemical soil conditions: Regional and temporal variability. (this volume)

Stinner, B. R., H. R. Krueger and D. A. McCartney. 1986. Insecticide and tillage effects on pest and non-pest arthropods in corn agroecosystems. Agric. Ecosystems Environ. 15:11-21.

Stinner, B. R., D. A. McCartney, and W. L. Rubink. 1984. Some observations on ecology of the stalk borer (Papaipema nebris (Gn.): Noctuidae) in no-tillage corn agroecosystems. J. Georgia Entomol. Soc. 19:229-234.

Stockdill, S. M. J. 1982. Effects of introduced earthworms on the productivity of New Zealand pastures. Pedobiologia 24:29-35.

Sumner, D. R., B. Doupnik, Jr. and M. G. Boosalis. 1981. Effect of reduced tillage and multiple cropping on plant diseases. Ann. Rev. Phytopathol. 19:167-187.

Thomas, G. W., R. L. Blevins, R. E. Phillips, and M. A. McMahon. 1973. Effect of a killed-sod mulch on nitrate movement and corn yield. Agron. J. 65:737-739.

Triplett, G. B. Jr., and D. M. Van Doren, Jr. 1969. Nitrogen, phosphorus, and potassium fertilization of non-tilled maize. Agron. J. 61:637-639.

Van Doren, D. M. Jr., G. B. Triplett, Jr., and J. E. Henry. 1977. Influence of long-term tillage and crop rotation combinations on crop yields and selected soil parameters for an Aeric Ochraqualf soil. Research Bull. No. 1091 of the Ohio Agric. Res. and Dev. Center, Wooster, Ohio.

CHAPTER 8

EFFECT OF CONSERVATION TILLAGE ON BIOLOGICAL AND CHEMICAL SOIL CONDITIONS: REGIONAL AND TEMPORAL VARIABILITY

M. S. Smith and R. L. Blevins,
University of Kentucky, Lexington, Kentucky

INTRODUCTION

In order to develop a better understanding of the management, environmental impact, and fundamental soil processes associated with conservation tillage it is natural and necessary that we seek generalizations about the differences between conservation tillage and conventional tillage systems. A sufficient number of observations indicate that some of these generalizations about conservation tillage effects on biological and chemical properties can be usefully and widely applied. An example, discussed in the previous paper, (Dick and Daniel, Chapter 7) is the concentration of nutrients, organic matter and microbial activity near the soil surface under reduced tillage.

On the other hand, it has become apparent that tillage (or the elimination of tillage) does not have the same effect at all times or at all locations. Quantitative and sometimes qualitative differences in the effects of conservation tillage were noted in the previous paper and will be emphasized here. Elsewhere, we and others have generalized about conservation tillage effects on chemical and biological properties (Blevins et al., 1983a; Blevins et al., 1983b; Blevins et al., 1977; Blevins et al., 1985; Phillips et al., 1980; Doran and Power, 1983; Unger and McCalla, 1980). Our emphasis on inconsistency here does not imply that we doubt the value of these generalizations.

We will present data illustrating that the effects of conservation tillage vary with: time of year, time (years)

Effects of Conservation Tillage on Groundwater Quality: Nitrates and Pesticides, Terry J. Logan et al., eds.

since initiation of the tillage system, climate, soil type and previous management. We can offer tentative explanations for some of these inconsistencies. However, in many cases we lack the ability to predict how tillage will interact with the parameters above and what will be the net effect on chemical and biological processes.

MICROBIAL POPULATIONS AND ACTIVITY

In most studies, increased numbers of microorganisms have been found at the soil surface as tillage is reduced. Doran (1980) enumerated several microbial types at seven locations in the U.S. where no-tillage vs. conventional tillage experiments are in progress. On the average, numbers of bacteria, fungi and actinomycetes counted were higher in no-till in the surface 7.5 cm of soil but higher in conventional tillage from 7.5 to 15 cm (Table 1). For the entire surface 15 cm, differences between tillage systems were small for most groups of organisms. In West Virginia, no tillage effects were observed on microbial populations in 0 to 30 cm samples (Fairchild and Staley, 1979). No attempt was made to measure differences in distribution within this depth of soil. In earlier Nebraska work, Dawson et al. (1948) and Norstadt and McCalla (1969), measured greater populations of microbes in the surface of stubble-mulched or sub-tilled soils compared to plowed soils.

Table 1. Ratio of Microbial Populations Between No-Tillage and Conventional Tillage, Average of Seven Locations (Doran, 1980).

Microbial group	Ratio NT/CT by depth (cm)		
	0-7.5	7.5-15	0-15
Fungi	1.4	0.7	1.0
Actinomycetes	1.1	1.0	1.1
Aerobic bacteria	1.4	0.7	1.0
NH_4 oxidizers	1.2	0.6	0.9
Denitrifiers	7.3	1.8	2.8

Some microbes are apparently more affected than others by tillage and residue management. Denitrifier and facultative anaerobic counts have been significantly greater in no-tillage than conventional tillage (Doran, 1980; Rice and Smith, unpublished data). This has led Doran to suggest that no-till is a less oxidative microbial habitat, that is anaerobic microbes and the processes they catalyze, such as denitrification, would generally be favored in no-till, while the opposite would be expected for aerobic organisms and processes, such as nitrification. Further direct process measurements are necessary to determine how widely this concept can be applied. On a well-drained soil in Kentucky, direct measurements of denitrification made at infrequent intervals did indicate a greater potential for denitrification in no-till (Rice and Smith, 1982). However, an N-15 balance study at the same site showed no significant tillage effect on N losses (Kitur et al., 1984). Still at the same location, nitrification rates were apparently greater under no-till when NH_4^+ availability was high (after fertilization), presumably because the moisture status near the soil surface was more favorable. Yet nitrification was apparently more rapid at other times in conventional tillage because of greater NH_4^+ availability (Rice and Smith, 1983).

The fumigation technique (Jenkinson and Powlson, 1976) of measuring microbial biomass C and N is being widely used to overcome many of the limitations of enumeration techniques based on isolation and growth of microbes. Doran (1987) has measured microbial biomass at multiple tillage experiments using this technique. On the average, these results are in agreement with the results obtained by plate counting. Biomass C was greater in the surface 7.5 cm of no-tillage soils, but below this depth was somewhat higher with conventional tillage. However, inspection of results at individual locations shows that this effect was not observed in every case (Fig. 1). Three possible patterns appear: at the Minnesota site there was little effect of tillage on biomass, at the Nebraska site the distribution of biomass was altered but the effect on total biomass in the profile was small, also at the Kentucky site the biomass was concentrated at the surface of no-tillage but in this case the total biomass to 30 cm was dramatically increased too.

It is relatively easy to explain the generally observed concentration of microbes near the surface with conservation tillage. This stratification phenomenon was discussed earlier (Dick and Daniels) and it was pointed out that the extent of stratification depends on the quantity of microbial substrate (plant residues) produced and on soil conditions. It is more difficult to understand why no-till should lead to a total increase in biomass in the profile if annual substrate input is approximately equal in both systems, as seems to be true at the Kentucky site.

It is important to recognize that most investigations of tillage effects on microbial processes and populations have used samples collected at a single time, or a small number of times,

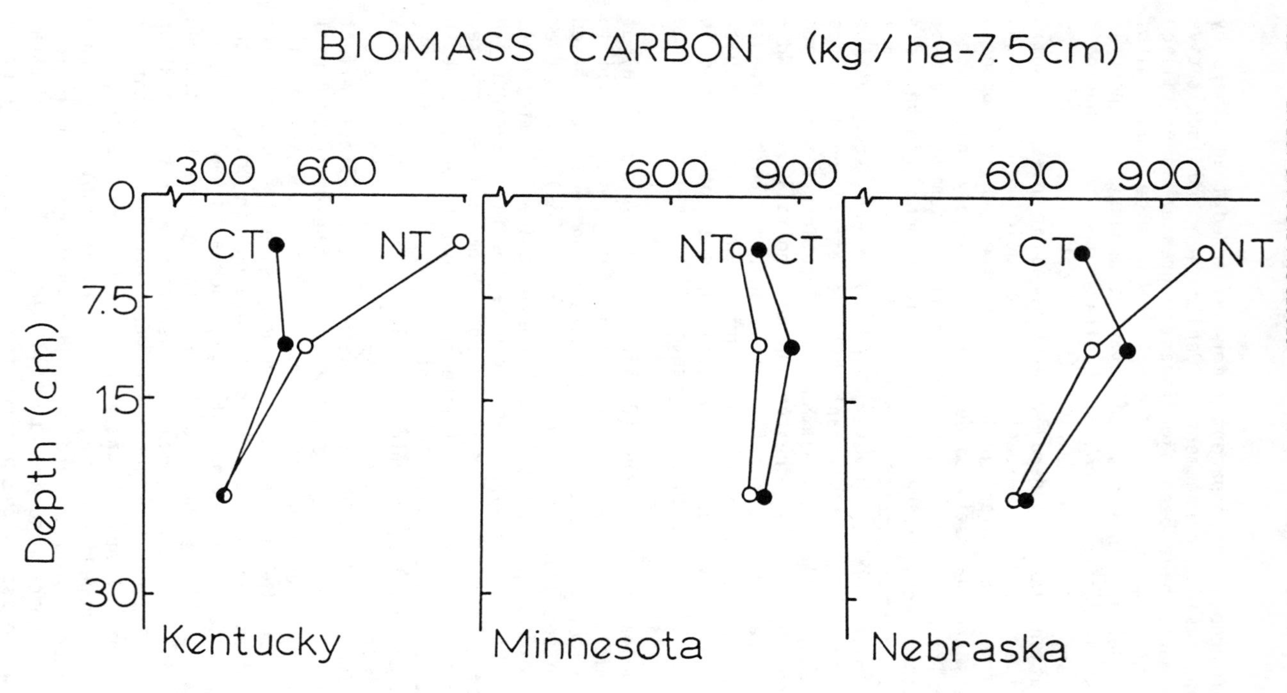

Figure 1. Microbial biomass carbon by depth for no-tillage and plowed soils at 3 locations (from Doran, 1987).

during the growing season. Yet is is clear that microbial activity varies significantly over the course of a year and it might be expected that the relationships among tillage systems would change. This effect is documented in Figure 2. Evolution of CO_2 was monitored under closed covers placed on Maury soil in Kentucky (Rice, 1983). In the spring microbial respiration was approximately equal in no-till and conventional tillage, perhaps because the effects of higher temperatures were nullified by the effects of lower moisture content in conventional tillage. Immediately after plowing and planting, respiration was significantly greater in conventional tillage, presumably because of an increase in substrate accessibility and temperature. Later in the season, respiration was greater in no-tillage. During this time there was little difference in temperature and the plowed soil was drier.

Table 2, adapted from data of Lynch and Panting (1980), shows that tillage effects on microbial biomass, as well as microbial activity, can vary seasonally. During most of the year, biomass was greater in the surface 5 cm of direct-drilled than in plowed soils, even though residues are burned off in these wheat production systems in England. One month after the plowed soil was disturbed, however, it had a higher biomass content than the undisturbed, direct-drilled system.

Table 2. Microbial Biomass in Direct-drilled and Plowed Wheat Soils (Lynch and Panting, 1980).

Time	Microbial biomass	
	DD	PLOW
	mg C /100g soil	
2 months before plowing	72	54
1 month after plowing	51	66
6 months after plowing	74	51
10 months after plowing	121	71

Integrated over an entire year, which system has greater rates of heterotrophic activity and carbon mineralization? It would be impossible or at least extraordinarily time-consuming to determine this by measuring CO_2 production, enzyme activities or microbial populations. The best answer may come from carbon budgets. For our long-term tillage experiment on the Maury soil in Kentucky, carbon inputs as plant residues have been similar, with a slightly greater quantity in conventional tillage, however soil carbon losses have been considerably greater in conventional tillage (see Fig. 1 in Dick and Daniels, Chapter

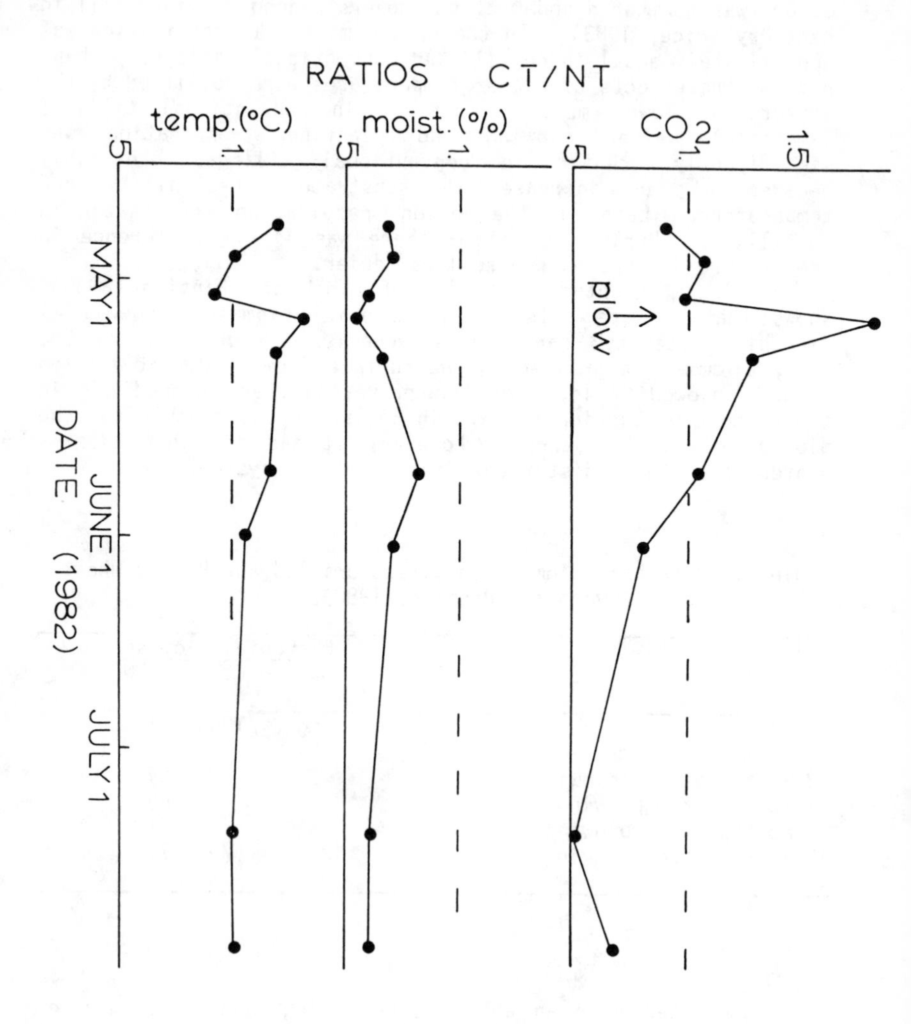

Figure 2. Ratio of conventional tillage to no-tillage temperature, soil moisture and CO_2 from the surface on various dates (from Rice, 1983).

VII). Since erosion losses are negligible for both systems at this site, the observations suggest that carbon mineralization and heterotrophic microbial activity over the long-term, have been greater in conventional tillage.

ORGANIC CARBON AND NITROGEN

As indicated in the previous paper, organic C and N generally accumulate at the soil surface under conservation tillage. This was further documented in a survey of 6 sites in the U.S. by Doran et al. (1985) (Table 3). These data indicate that organic C and biomass C were increased by 40 and 58%, respectively, in the surface 7.5 cm of no-tillage soil. Below this depth there was little difference. Therefore, the total organic matter in the soil was relatively increased with no-tillage.

This difference is not observed universally as shown by the data of Carter and Rennie (1982) for 4 Canadian soils (Table 3). At these locations there was no increase and perhaps a slight decrease in organic C in the zero till soils. The different tillage systems had been imposed on these soils for 2 to 16 years, depending on location, so it is difficult to argue that insufficient time had elapsed for tillage effects to become apparent.

Table 3. Effect of Tillage on Soil Organic Matter and Microbial Biomass at Multiple Sites in the U.S. and Canada (Doran et al., 1985; Carter and Rennie, 1982).

Soil depth (cm)	Relative difference, NT vs Ct	
	Organic C	Biomass C
	%	
	U.S. (6 sites)	
0 - 7.05	+40	+58
7.5 - 15.0	- 1	- 2
	Canada (4 sites)	
0 - 5	-11	+11
5 - 10	- 2	+ 6

One important factor in accounting for regional differences may be the level of residue input. In the Canadian study this was apparently in the range of 2.2 to 4.5 Mg dry matter ha/yr, derived solely from the wheat. In Kentucky at the Maury site, which shows pronounced soil organic matter differences, corn stover inputs alone are generally in the range of 6 to 9 Mg ha/yr (Kitur, 1982). The winter cover crop provides approximately 3 Mg/ha additional dry matter as residue. Therefore, residue input can vary by as much as a factor of 3 to 4 from one location to another. It would be expected that effects of tillage on soil C would be more dramatic as residue rate increased.

It might be predicted that organic matter would accumulate under conservation tillage to a lesser extent in warm climates than in cold. This pattern is not readily apparent in the literature, as seen in Table 3. One confounding factor is the greater potential for winter cover crop production in warm climates. We also speculate that more rapid water evaporation in warm climates could dessicate organic materials at the soil surface and greatly retard degradation. This effect could be less significant for residues incorporated into the soil by conventional tillage.

The Canadian soils were in conventional tillage prior to initiation of the zero tillage experiments and this may also minimize differences in organic carbon accumulation. As shown in Table 4, previous management can be very important in determining the magnitude of conservation tillage effects (Doran and Power, 1983). Two long term tillage experiments were conducted in Nebraska, one begun on a native sod, the other on a previously cultivated soil. It is apparently easier to degrade than to augment soil organic matter; the rates of organic N loss from the sod exceeded the rates of gain in the previously cultivated soil. In both cases the no-till soil contained more total N than the plowed soil after 9 to 10 years. However, the difference in annual losses or gains due to tillage was only 7 kg N/ha/yr for the previously cultivated experiment and was 42 kg N ha/yr for the native sod site.

AVAILABLE SOIL NITROGEN

It is commonly observed that NO_3^- concentrations are higher in plowed soils than in soils with reduced tillage (NaNagara et al., 1976; Dowdell and Cannell, 1975; House et al., 1984). This difference is illustrated in Figure 3 with data from a Kentucky study by Thomas et al. (1973). Nitrate concentration is an important parameter not only as related to crop nutrition but also as related to N transport to groundwater. Soil NO_3^- concentrations could be lower because of slower production, through mineralization and nitrification, or because of greater

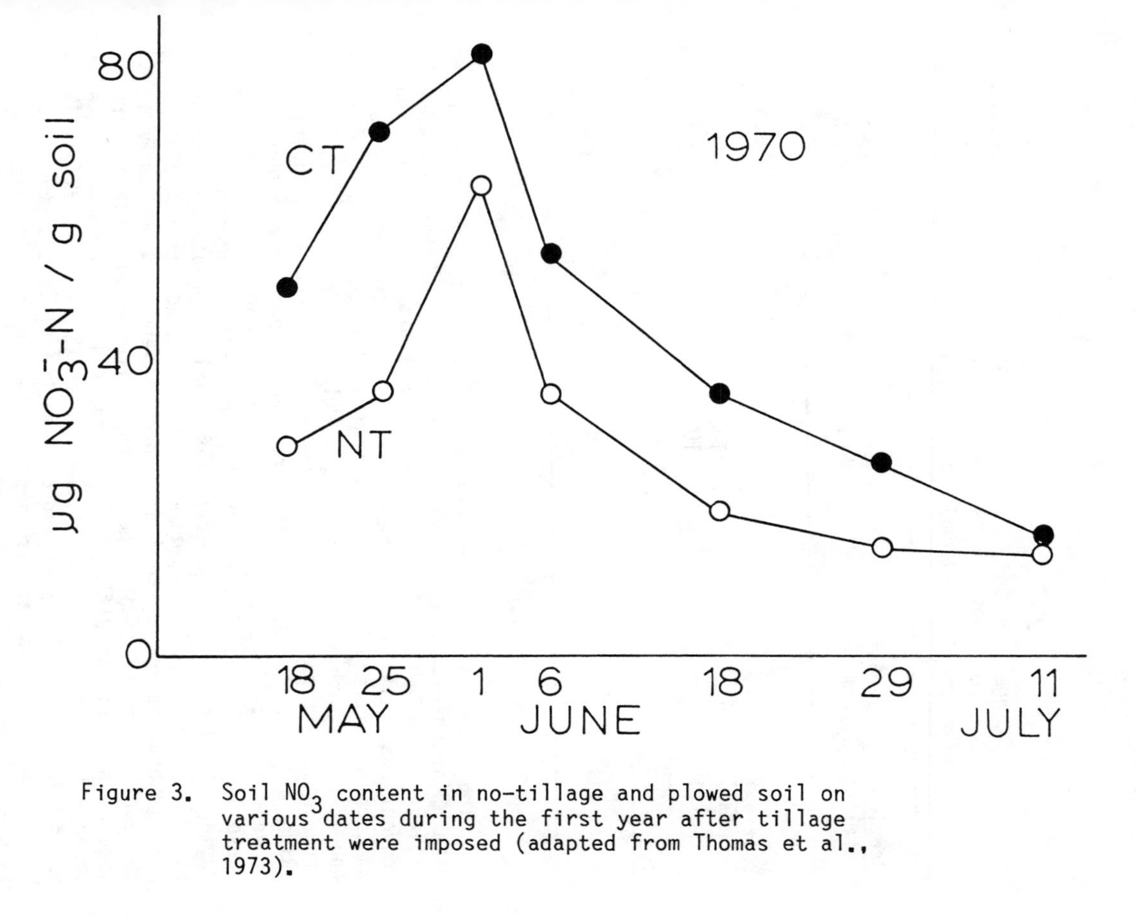

Figure 3. Soil NO_3 content in no-tillage and plowed soil on various dates during the first year after tillage treatment were imposed (adapted from Thomas et al., 1973).

consumption and losses, through immobilization or leaching or denitrification.

Table 4. Changes in Soil Nitrogen (0-30 cm) for Wheat-fallow Systems with Differing Tillage and Previous Management. (Doran and Power, 1983).

Previous Management Tillage System Imposed	Time of Tillage Study	Change in Soil N
	yrs	kg/N/ha/yr
Cultivated		
Plow	10	+16
No-till	10	+23
Native sod		
Plow	9	-109
No-till	9	- 67

The observations in Figure 3, and most of the literature data on inorganic N, were made soon after the initiation of the tillage experiment. After 10 to 15 years of no-till and conventional tillage at the same site, a somewhat different pattern of tillage effects on N availability is emerging. In the 12th and 13th years inorganic N (NH_4^+ plus NO_3^-) was measured frequently and no consistent effects of tillage were observed (Rice et al., 1986). Nitrate concentrations were higher in conventional tillage only during the early part of the growing season (Fig. 4), but NH_4^+ was greater in no-till. At other times during the year, NO_3^- in no-till was equal or greater. Lamb et al. (1985) have also reported that increased inorganic N in plowed wheat-fallow systems in Nebraska were observed only in the early years of their experiments. After 2 to 6 years there were no significant differences between tillage systems. These observations indicate that tillage effects on soil NO_3^- will depend on both time of year and on time the tillage systems have been in effect. The same is likely to be true for NO_3^- leaching. In many climates, including ours, leaching losses are greatest in late winter and early spring. Tillage effects observed during the summer growing season cannot always be extrapolated to other times of the year.

A related illustration of the importance of considering long-term tillage effects is given in Figure 5. In the first 5 years of this Kentucky experiment corn yields were consistently greater with conventional tillage than with no-tillage at low N

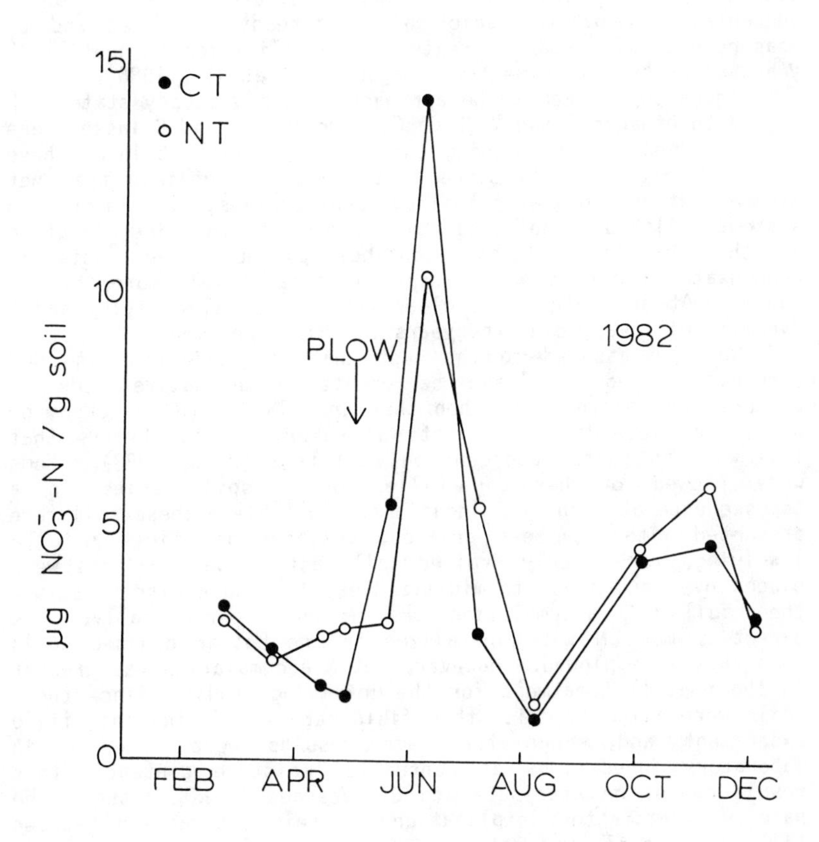

Figure 4. Soil NO_3^- content in no-tillage and conventional tillage by date during the thirteenth year after tillage treatments were imposed (from Rice, 1983).

fertilizer rates. This effect has been observed at many locations (Meisinger et al., 1985: Triplett and Van Doren, 1969; Moschler and Martens, 1975; Legg et al., 1979). We attribute this primarily to greater N mineralization when this soil, previously in bluegrass sod, was plowed. Differences in leaching, denitrification and immobilization may also be involved. In the last 5 years (no yield was obtained in 1983 due to drought) yields at low N rates have not been significantly affected by tillage (Rice et al., 1986). Apparently, the no-till soil is now supplying as much N as the plowed soil. This is further indicated by the equal concentrations of inorganic soil N already discussed and by measurements of equal or greater N mineralization in no-till 13 years after the experiment was begun (Rice et al., 1986).

These soils seem to be approaching a new steady state with regard to organic C and N (Fig. 6). Organic N and C losses were rapid in both systems during the first 5 years but levels have been relatively stable since that time. We believe that net mineralization and N availability have become similar in the two systems. Although the specific rate constant for mineralization in the no-till soil may continue to be lower, this is compensated for by the accumulation of relatively more total C and N. At any rate it is clear that microbial activity and N dynamics will vary over the years in this experiment.

We have assumed in the past that mineralization and NO_3^- accumulation would always be greater when native sods or pastures are plowed than when they are planted to row crops by no-tillage methods. Thus, it was surprising to observe that this was, in fact, dependent on soil type (Rice, 1983). Sods were plowed or herbicide-killed on 3 soil series in a toposequence at each of 2 locations. In Table 5 these soils are presented with the best drained, upslope soil first and the low-lying, more poorly drained soil last. Open shelters were placed over sub-plots to minimize leaching and denitrification, then soil NO_3^- accumulation was measured periodically. As expected, more N was mineralized in the better drained soils when they were plowed. However, net N accumulation was greater in the no-till treatment for the low-lying soils. Since these soils were slightly drier than field capacity during this field experiment and since the same results were observed in laboratory incubations at controlled moisture contents, this result cannot be attributed to differences in NO_3^- losses. We have no satisfactory explanation for this interaction between tillage and soil type determining net N mineralization. It is offered as another example that conservation tillage will not have the same effects at all locations.

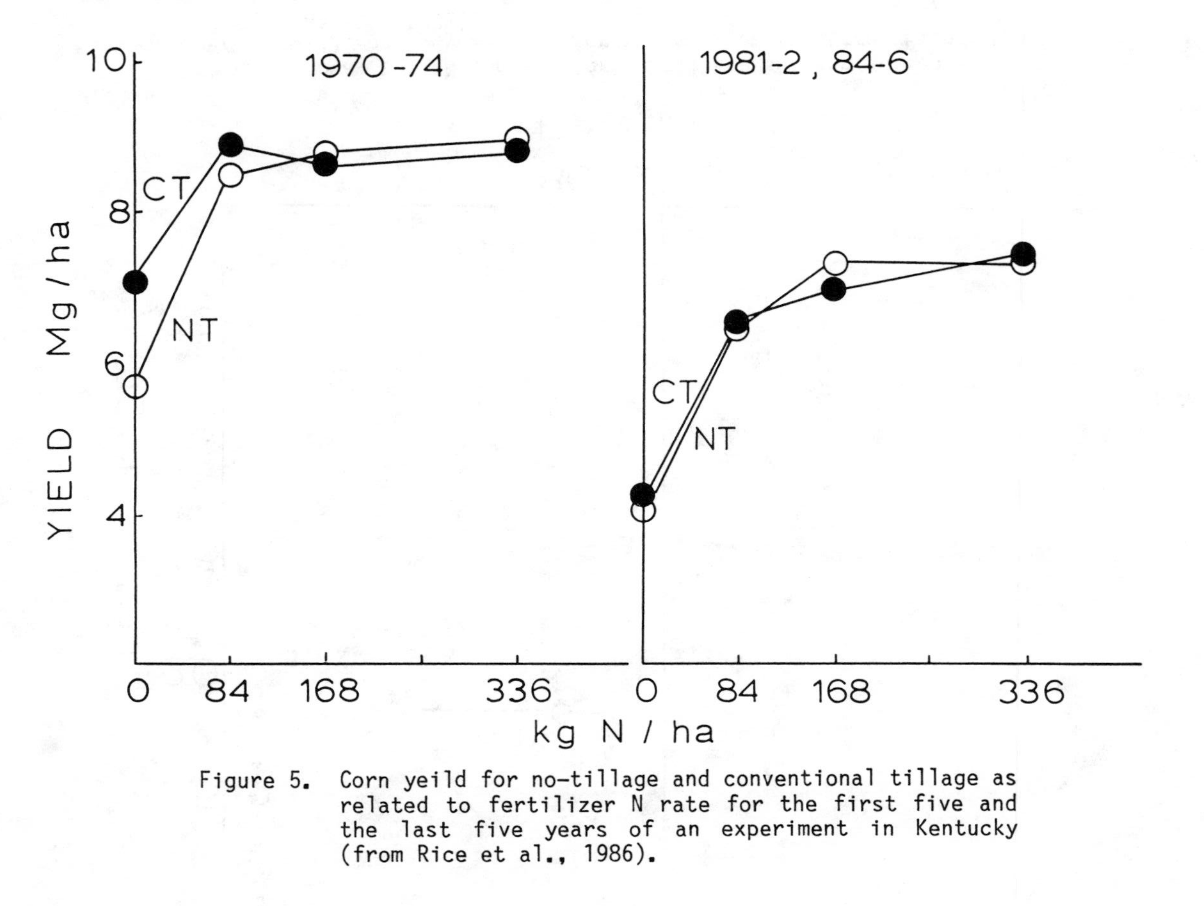

Figure 5. Corn yeild for no-tillage and conventional tillage as related to fertilizer N rate for the first five and the last five years of an experiment in Kentucky (from Rice et al., 1986).

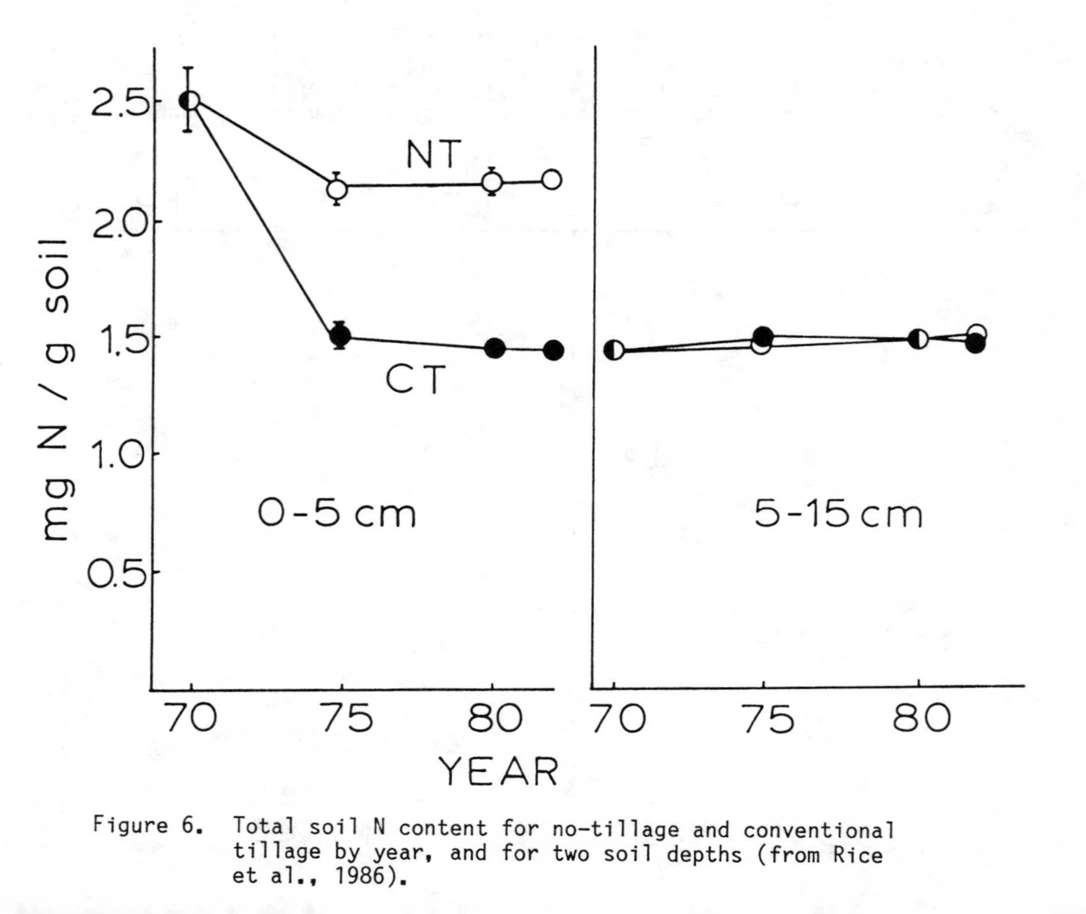

Figure 6. Total soil N content for no-tillage and conventional tillage by year, and for two soil depths (from Rice et al., 1986).

Table 5. Amount of N Mineralization in the Field for No-till (NT) and Plowed (CT) Treatments on Several Soils at Two Locations (During 2 months after tillage), (Rice, 1983).

Location	Soil	N mineralized	
		NT	CT
		ug N/g soil	
Lexington	Maury	40	92
	Lindside	73	84
	Lanton	48	29
Princeton	Zanesville	17	22
	Johnsburg	9	12
	Purdy	11	8

SUMMARY

The commonly observed effects of conservation tillage on chemical and biological properties of soil have been discussed more completely in the preceding paper and elsewhere. The most widely applicable generalization is that conservation tillage results in a stratification of the soil, with a relative accumulation of nutrients, acidity, organic matter and microbes near the surface. Our purpose has been to emphasize that the magnitude of this effect, and the nature of other biological and chemical effects, will vary. The effects of any tillage system will depend on climate, soil type, quantity of residues produced, previous management history, time of year and time since the tillage system was initiated. We do not expect that reduced evaporation by surface residues will have the same effect on microbial activity in a poorly-drained clay and in a well-drained sand. Reduced temperatures under a no-till mulch can decrease microbial respiration in Minnesota but increase it in Nigeria. Because tillage *per se* is not a primary, direct determinant of biological activity, it seems that predicing tillage system effects on biological activity must be an indirect, two-step process. First, we must understand how tillage systems influence the chemical and physical enviroment of soil organisms, particularly temperature, moisture and substrate availability. Second, we must understand, in a quantitative and dynamic way, how microbes and the reactions they catalyze are affected by these environmental changes.

REFERENCES

Blevins, R. L., W. W. Frye and M. S. Smith. 1985. The effects of conservation tillage on soil properties. In F.M. D'Itri (ed.) A systems Approach to Conservation Tillage. Lewis Publishers Co., Chelsea, MI.

Blevins, R. L., G. W. Thomas, M. S. Smith, W. W. Frye and P. L. Cornelius. 1983a. Changes in soil properties after 10 years continuous non-tilled and conventionally tilled corn. Soil and Tillage Research 3:135-146.

Blevins, R. L., M. S. Smith, G. W. Thomas and W. W. Frye. 1983b. Influence of conservation tillage on soil properties. J. Soil Water Conserv. 38:301-305.

Blevins, R. L., G. W. Thomas and P. L. Cornelius. 1977. Influence of no-tillage and nitrogen fertilization on certain soil properties after 5 years of continuous corn. Agronomy Journal 69:383-386.

Carter, M. R. and D. A. Rennie. 1982. Changes in soil quality under zero tillage farming systems: Distribution of microbial biomass and mineralizable C and N potentials. Can. J. Soil Sci. 62:587-597.

Dawson, R. C., V. T. Dawson and T. M. McCall 1948. Distribution of microorganisms in the soil as affected by plowing and subtilling crop residues. Research Bulletin 155, University of Nebrasks, Lincoln.

Doran, J. W. 1980. Soil microbial and biochemical changes associated with reduced tillage. Soil Sci. Soc. Am. J. 44:765-771.

Doran, J. W. 1987. Microbial biomass and mineralizable nitrogen distribution in no-tillage and plowerd soils. Biology and Fertility of Soils (in press).

Doran, J. W. and J. F. Power. 1983. The effects of tillage on the nitrogen cycle in corn and wheat production. In R. R. Lowrence (ed.) Nutrient Cycling in Agricultural Ecosystems. The University of Georgia, Spec Pub. 23.

Doran, J. W., L. N. Mielke and J. F. Power. 1985. Tillage imposed changes in the agricultural ecosystem. In Proc. 10th Conf. Intl. Soil Tillage Res. Org. Guelph, Ontario.

Dowdell, R. J. and R. Q. Cannell. 1975. Effect of plowing and direct drilling on soil nitrate concent. J. Soil Sci. 26:53-61.

Fairchild, D. M. and T. E. Staley. 1979. Tillage method effects on soil microbiota and C/N reservoirs. Abstracts of the American Society of Agronomy. 1979. Fort Colline, Colorado, p. 157.

House, G. F., B. R. Stinner, D. A. Crossley, Dr., E. P. Odum and G. W. Langdale. 1984. Nitrogen cycling in conventional and no-tillage agroecoystems. J. Soil Water Conserv. 39:194-200.

Jenkinson, D. S. and D. S. Powlson. 1976. The effect of biocidal treatments on metabolism in soil. V. A method of measuring soil biomass. Soil Biol. Biochem. 8:209-213.

Kitur, K., M. S. Smith, R. L. Blevins and W. W. Frye. 1984. Fate of ^{15}N depleted ammonium nitrate applied to no-tillage and conventional tillage corn. Agron. J. 76:240-242.

Kitur, B. K. 1982. Fate of nitrogen in no-tillage and conventional tillage corn systems using labeled NH_4 NO_3. M. S. Thesis. University of Kentucky, Lexington.

Lamb, D. A., G. A. Peterson and C. R. Fenster. 1985. Fallow nitrate accumulation in a wheat - fallow rotation as affected by tillage system. Soil Sci. Soc. Am. J. 49:1441-1446.

Legg, J. O., G. Stanford and O. L. Bennett. 1979. Utilization of labeled N fertilizer by silage corn under conventional and no-till culture. Agron. J. 71:1009-1015.

Lynch, J. M., and L. M. Panting. 1980. Cultivation and the soil biomass. Soil Biol. and Biochem. 12:29-33.

Meisinger, J. J., V. A. Bandel, G. Stanford and J. O. Legg. 1985. Nitrogen utilization of corn under minimal tillage and moldboard plow tillage. I. Four year results using labeled N fertilizer on Atlantic Coastal Plain Soil. Agron J. 77:602-611.

Moschler, N. W. and D. C. Martens. 1975. Nitrogen, phosphorus and potassium requirement in no-tillage and conventionally tilled corn. Soil Sci. Soc. Am. Proc. 39:886-891.

NaNagara, T., R. E. Phillips and J. E. Leggett. 1976. Diffusion and mass flow of nitrate nitrogen into corn roots grown under field conditions. Agronomy J. 68:67-72. also 68:63-67.

Norstadt, F. A., and T. M. McCalla. 1969. Microbial populations in stubble-mulched soil. Soil Sci. 107:188-193.

Phillips, R. E., R. L. Blevins, G. W. Thomas, W. W. Frye and S. H. Phillips. 1980. No-tillage agriculture. Science 208:1108-1113.

Rice, C. W., and M. S. Smith. 1982. Denitrification in no-till and plowed soils. Soil Sci. Soc. Am. J. 46:1168-1173.

Rice, C. W., and M. S. Smith. 1983. Nitrification of fertilizer and mineralized ammonium in no-till and plowed soil. Soil Sci. Soc. Am. J. 47:1125-1129.

Rice, C. W. 1983. Microbial nitrogen transformating in no-till soils. Ph.D. Thesis. University of Kentucky, Lexington.

Rice, C. W., M. S. Smith and R. L. Blevins. 1986. Soil nitrogen and availability after long-term continuous no-tillage and conventional tillage corn production. Soil Sci. Soc. Am. J. 50:1206-1210.

Thomas, G. W., R. L. Blevins, R. E. Phillips and M. A. McMahon. 1973. Effect of a killed sod mulch on nitrate movement and corn yield. Agronomy J. 65:736-739.

Triplett, G. B., Jr. and D. M. VanDoren, Jr. 1969. Nitrogen, phosphorus and potassium fertilization of non-tilled maize. Agron. J. 61:637-639.

Unger, P. W. and T. M. McCalla. 1980. Conservation tillage systems. Adv. Agron. 33:1-58.

SECTION III

EFFECT OF CONSERVATION TILLAGE SYSTEMS ON FATE AND TRANSPORT OF APPLIED PESTICIDES AND NITROGEN FERTILIZER

CHAPTER 9

THE EFFECTS OF CONSERVATION TILLAGE PRACTICES ON PESTICIDE VOLATILIZATION AND DEGRADATION

D. E. Glotfelty,
USDA-ARS, Beltsville, Maryland

INTRODUCTION

"Conservation Tillage" is a generic term encompassing many different soil management practices that vary both in the amount of tilling that is actually done to the soil, and in the amount of plant residue that is left on the soil surface. For this discussion, I will not attempt to define and discuss all these different practices, but will rather compare the effects upon pesticide dissipation of the extremes of these practices: conventional seed bed preparation through plowing and discing, and no-till, in which the seed is introduced into unplowed soil in narrow openings sufficiently deep to cover the seed and provide soil contact. Compared to conventional till, no-till changes soil properties and soil surface mircoclimate in ways that may profoundly affect pesticide dissipation by volatilization, degradation, and plant uptake because these dissipation processes are influenced by soil temperature, soil-water content, soil-organic matter, soil water and air movement, and soil chemical properties such as pH.

EFFECTS OF NO-TILL ON SOIL PROPERTIES

The effects of no-tillage agriculture upon soil properties have been extensively studied (Phillips and Phillips, 1984). In the no-till system, a mulch of plant material is present at the soil surface. The amount of surface residue varies as does its condition; for example, residues may be an actively growing

Effects of Conservation Tillage on Groundwater Quality: Nitrates and Pesticides, Terry J. Logan et al., eds.

vegetative cover, or the decaying residue of the previous crop. Management of this plant residue affects the extent to which it influences pesticide dissipation. It is obvious this surface mulch will intercept part of the pesticide spray and thus interfere with surface coverage. This is one reason more pesticide usually is required to control pests in no-till agriculture. If rain occurs soon after application, part of the pesticide will be washed from the mulch onto the soil, but usually a significant amount remains in the mulch. The dissipation of this part will be different from that which reaches the soil.

The change in the microclimate that occurs at the soil surface when a plant residue mulch is present changes the energy balance, the moisture distribution, and the rate of gas exchange at the soil surface. These all have a significant effect on pesticide dissipation. The mulch intercepts sunlight, and due to its greater albedo, more of the radiation is reflected back to the sky. Therefore, less sunlight energy is available at the soil surface. The mulch layer also insulates the soil surface so that conduction of heat into and out of the soil is reduced. Diurnal temperature changes are, therefore, less pronounced, and, on the average, no-till soils are 2° to 10°C cooler than bare soil, especially when the soil is warming up in the spring. Although gas exchange within the mulch layer may be rapid due to the increased aero-dynamic roughness of the mulch, gas exchange and diffusion from the soil surface is reduced compared to bare soil. Cooler temperatures, lower heat conduction, and slower vapor diffusion mean that water loss from mulched no-till soil is slower than from a bare soil surface, and may lead to water saturation and localized anaerobic conditions in wet climates. Thus, no-till will cause a profound change in the soil water regime, particularly at the soil surface. The availability and distribution of soil water greatly influence the rate of pesticide dissipation.

Another very significant change is the amount and distribution of soil organic matter. In conventional-tillage systems, the organic matter is uniformly distributed throughout the plow depth. No-till systems return plant residues to the soil surface where it gradually decomposes. This adds to the amount of soil organic matter, and concentrates it in a shallow zone at the soil surface in direct contact with the applied pesticides. Organic matter is another very important variable in pesticide dissipation.

The change in soil structure and bulk density that accompanies no-till farming changes the nature of water and air movement through diffusion and mass flow within the soil. Untilled soil, partly through the increased organic matter content, has improved soil aggregate stability and increased aggregate size. It also has more macropores and biochannels, the continuity of which are not disturbed by plowing. These changes in soil structure generally favor better mass movement of soil water and air, but the higher bulk density and volumetric water contents of untilled soil results in lower air-filled porosity. There may therefore be areas of relatively immobile soil water and localized anaerobic conditions, especially in fine textured soils.

Finally, conservation tillage changes soil chemical properties, expecially at the soil surface. This is because most fertilizer, lime, and other agricultural chemicals are applied to the surface, and these amendments are not mixed into the soil. They only move into the soil if they do so with water. Organic matter, P, and K increase markedly near the surface, but deeper soil may become depleted in nutrients. The addition of ammoniacal nitrogen fertilizer causes a loss of Ca and Mg from the surface and an increase of exchangeable Al and Mn, resulting in a dramatic drop in surface soil pH. This surface acidification is a primary change in soil chemistry under no-till systems, and may exert a large effect on pesticide dissipation.

PESTICIDE VOLATILIZATION

Research over the past two decades has shown that the quantity of pesticide that volatilizes is usually much larger than that which moves with runoff or leaching and, for most pesticides, is second only to degradation in causing dissipation from a treated field (Taylor, 1978). Important factors that control pesticide volatilization include: the intrinsic physicochemical properties of the pesticide (vapor pressure is obviously important); how the pesticide is formulated, applied, and managed in the field; the energy input and mass transfer processes within the soil; and the properties and condition of the soil, especially organic matter and water contents. With the exception of the intrinsic pesticide properties, all of these factors may be very different in no-tillage systems compared to conventional systems.

Pesticide management can control volatility. The most important variable controlling pesticide volatilization is whether the pesticide is incorporated into the soil or allowed to remain on the surface. Even shallow incorporation can produce a many-fold decrease in volatility (Taylor, 1978). In no-till systems, pesticides usually remain on the surface and are thus potentially volatile, but sometimes they are incorporated by banding into the seed row. Pesticide formulation and mode of application can influence volatility, but the effects are usually only of secondary importance, unless the formulation is especially designed to restrict volatilization such as by microencapsulation (Turner et al., 1978). Potential volatility problems in conservation tillage systems may make such formulations more attractive. Vegetable oil formulations also help control volatilization, compared to emulsifable-concentrate formulations. Wettable-powder formulations will erode with the wind and carry away even non-volatile pesticides. However, such wind erosion losses appear to be small.

The adsorption of pesticide to the soil is also important; the more strongly the pesticide is adsorbed, the lower will be its volatility. Water competes for absorption sites and many pesticides are strongly but reversibly adsorbed to dry soil (Spencer and Claith, 1974). For this reason, the third most important factor determining volatility from soil, after vapor pressure and soil incorporation, is the soil-moisture content and the manner in which the moisture is distributed. A dry layer of soil at the surface can greatly restrict volatilization even of incorporated pesticides (Glotfelty et al., 1984).

Volatilization from moist bare soil peaks around solar noon, and falls to much lower values at night. This diurnal pattern, which can produce marked differences in volatilization between day and night, comes about because sunlight provides the soil/atmosphere system with the energy needed to cause volatility to increase. Soil temperature rises, providing latent heat of evaporation, which causes an increase of pesticide vapor density at the soil surface. Pesticides released from the surface are rapidly diluted and carried away be eddy dispersion-turbulent atmospheric mixing that is most intense at noon and extends to within a few mm of the bare soil surface. At night, the soil cools off, turbulence dies away, and volatilization subsides. Changing these energy balance and gas exchange relationships will result in a change in the rate of volatilization.

The rate of volatilization of incorporated pesticides, or those that are washed into the soil, is controlled by their movement throught the soil to the soil surface. This occurs by two processes: diffusion-controlled flow, and convective flow (Jury et al., 1984). The rate of diffusion-controlled flow depends upon the overall diffusion coefficient of the pesticide in the particular soil of interest, which in turn depends upon the temperature and air-filled porosity. The latter depends upon the bulk density and liquid water content. Diffusion-controlled flow is more effective for low water solubility, high vapor pressure pesticides. Convective flow, on the other hand, occurs when evaporation at the surface induces upward water flow. The upward-flowing water carries dissolved pesticide to the surface. Convective flow will be more important for pesticides with higher water solubility.

When we compare how the factors that control pesticide volatilization are different between conventional and no-tillage systems, we find offsetting effects: some factors tend to increase volatilization, while others tend to reduce it. For this reason, the net effect of all the changes may be difficult to predict and careful field experimentation will likely be required to evaluate the over-all effects of tillage.

To the extent that conservation tillage promotes unicorporated, surface-applied pesticides more than conventional tillage, it has the potential for greatly increasing pesticide volatilization. This may be especially true for that part of the spray that is intercepted and remains in the surface mulch. Studies of pesticide volatilization from plant surfaces indicate that pesticides are less strongly adsorbed to vegetation than they are to soil (Willis et al., 1983). The pesticide held up in the mulch has a larger surface area exposed to the wind. Finally, the vegetative mulch is aerodynamically rougher than bare soil so that better gas exchange occurs within the mulch layer than at the soil surface. For these reasons the pesticide intercepted by the mulch may be lost very rapidly.

That part of the spray that penetrates the surface mulch and is adsorbed at the soil surface is in a protected microenvironment, but is still potentially volatile. All other factors being equal, the loss from conservation tillage will likely be slowed somewhat compared to bare soil because of the reduction in gas exchange at the soil surface caused by the presence of the mulch layer. Pesticide vapor at the soil surface of a conservation tillage system must diffuse further into the mulch layer before incipient turbulent transport becomes operative. By contrast, pesticide vapor at the soil surface of a plowed system experiences an effective laminar layer, or diffusion path, of only a few mm.

The increased organic matter content at the soil surface in conservation tillage further reduces volatilization losses because of increased pesticide adsorption. Research has shown that doubling the soil organic matter content cuts the volatilization rate roughly in half (Spencer and Claith, 1974). Finally, the 2°-to-10°C cooler soil temperatures typical for conservation tillage will also significantly reduce volatilization losses, perhaps, by as much as a factor of two.

These reductions in volatilization from conservation tillage systems may be offset by the effect of the increased moisture content and altered moisture distribution. Many plowed soils tend to readily form a very dry surface layer. This dry layer reduces volatilization by as much as an order of magnitude or more compared to moist soil (Glotfelty et al., 1984). If, on the other hand, the soil surface under the mulch layer of a no-till system remains moist for a longer period, volatilization remains at high levels longer, and losses may be correspondingly larger.

Pesticides that are present much below the soil surface, either by incorporation or "wash in," are probably less volatile in conservation tillage systems. The presence of biochannels and macropores increases air and water flow, and thus ostensibly increases diffusion and mass flow of pesticides in no-till soils. However, it is not clear to what extent these macropores affect the whole soil mass. Away from these preferential-flow pathways there may be stagnant volumes where little exchange of air and water occur. The lower air-filled porosity and lower temperature of no-till systems imply general restriction of diffusive flow. The lower evaporation of water implies a reduction also in mass flow through the soil. On balance, these factors seem to favor reduced pesticide volatility with no-till. On the other hand, volatilization of incorporated pesticides, even in conventionally-tilled soils, is very low, so that differences between the two systems will also be small, and perhaps insignificant in terms of the total mass balance.

PESTICIDE DEGRADATION

The changes in chemical and physical properties of soils that occur in no-tillage systems may greatly affect pesticide degradation rates. For example, surface acidification occurs unless the soil is periodically limed. The soil pH has a marked effect on pesticide degradation rates. Chemical reactions may be directly affected by pH changes, as when hydrolysis is catalyzed by either an acid or base. Such reactions may have sharp maxima or minima, and hence a small change in soil pH may produce a many-fold change in reaction rate (Kells, et al., 1980).

The microbial population and the types of organisms active in pesticide degradation also change with pH change. In the pH range of 7.5 to 8, actinomycetes are dominant; at pH 6 to 7.5, bacteria; and at low pH, and 5, soil fungi are dominant. Thus, fungi may be more important in conservation tillage systems. While the different soil microorganisms generally degrade pesticides by similar mechanisms, fungi are better at degrading adsorbed pesticides, and at causing dealkylation reactions.

Since both microbial population and chemical reactions change with soil pH, the relative importance of microbial vs. chemical degradation may also change as pH changes. This will influence the types of degradation products formed. Furthermore, these pH effects are very complex and are largely unpredictable and thus need to be experimentally determined (Hamaker, 1972).

The high concentration of pesticide at the soil surface may also influence the rate of degradation. Usually, as concentration increases, the specific degradation rate decreases, that is, a smaller percent of the initial concentration is degraded per unit time (Hamaker, 1972). This may lead to problems of pesticide carry over into the next season. On the other hand, high surface concentrations may produce a microenvironment favorable to degrading microbes and thus cause an accelerated rate of degradation, such as found in "problem soils" where the microorganisms have been selected which are adapted to breaking down certain classes of compounds. This effect may influence the breakdown of chemicals used the next growing season, although there is no evidence yet that the "problem soil" phenomenon is affected by tillage practice (Donald Kaufman, USDA-ARS, Soil Microbial Systems Laboratory, Beltsville, MD. Personal Communication).

The increase of organic matter in the surface soil will favor both greater microbial activity and stronger adsorption. The organic matter supports a higher microbial population and may, therefore, result in accelerated pesticide breakdown. However, the increased adsorption onto the organic matter may shield the pesticide from microbial attack resulting in a decreased rate of degradation, unless it also happens that chemical degradation increases upon adsorption, such as by surface catalysis.

The changes in water content and temperature that occur in no-tillage will also affect chemical and microbial degradation. The activation energies for pesticide degradation in soil range from about 3 to 30 kcal/mole. Therefore, reaction rates will be cut by as much as half for each 5° to $10^{o}C$ drop in temperature. This suggests that degradation rates in the cooler conservation tillage soil will be significantly slower, all other factors being equal. Degradation rates are much slower in dry soil than in moist, which suggests faster rates in moister no-tillage soil. However, the increase in rate of degradation tends to level off with higher moisture contents, and it is not clear whether the moisture effect will be significant. Finally, anaerobic sites may occur in no-till soils under wet conditions. Pesticide degradation in these sites will have entirely different reaction mechanisms, products, and kinetics compared to aerobic degradation.

A final uncertainty has to do with the degradation of that part of the pesticide application that is intercepted by the surface mulch. How the rate of degradation of this part compares to soil degradation is largely unknown. There has been some conjecture that the plant residue is degraded largely by fungi, and that any pesticide within that organic material will be degraded in like manner. Airborne spores are much more important than fungi in the soil for decomposing plant residues in the mulch, but no one has ever shown whether this particular group of organisms function to degrade pesticides.

CONCLUSION

Conservation tillage changes soil properties, soil chemistry, and the microclimate at the soil surface in ways that may profoundly affect the volatilization, degradation, and uptake of pesticides. Many of these changes would have predictable effects on pesticide dissipation, were they to occur in a homogeneous system such as plowed soil. We know, for example, the effects of soil moisture, organic matter, and temperature on pesticide volatilization and degradation. However, no-till systems are very heterogeneous and many of the changes that occur in no-till will have offsetting effects on pesticide dissipation. Because these systems have not been extensively investigated, the end result is frequently difficult to predict. It is clear that careful experimentation will be required to determine the overall effects of conservation tillage on pesticide dissipation.

One of the most significant changes that may accompany conversion to conservation tillage is in how the pesticide is managed. If a pesticide that is normally incorporated into the soil in a conventional tillage system is instead applied to the surface without incorporation in a no-till system, it is likely that a significant change in the rate and mechanisms of dissipation will occur. A second significant difference between conventional and no-till systems is the presence of the surface mulch that intercepts the pesticide spray and alters the microclimate at the soil surface. Other important differences include soil temperature, altered moisture content and distribution within the soil, and altered organic matter content and distribution within the soil.

Changes in pesticide degradation rates and products are almost certain to accompany the very different conditions in no-till soils. Even small changes in degradation rates may be very important. For example, doubling the half life of a herbicide may create a severe carryover problem affecting next year's crop. Changing degradation rates and products could also affect the environmental impact of a particular pesticide. Some degradation products are not only biologically active but are much more mobile than the parent pesticide. The nature of these changes in degradation rates and products have not been systematically studied for many pesticides.

One may conclude that volatilization is potentially much greater for some pesticides in no-till systems, resulting in both loss of efficacy and lower loading to waterborne transport. However, volatilized pesticides do return to the surface. Because of their relatively large ratio of surface area to watershed area, pesticide loading to the Great Lakes could conceivably increase following extensive adoption of no-till agriculture, because rainout ad dry deposition of atmospheric pesticides could be greater than waterborne loading from conventional tillage agriculture.

It should be pointed out that this discussion has focused in the extremes of no-till *versus* conventional-tillage systems. Some of the other conservation tillage practices that do till the soil to a certain extent and leave much less surface mulch than no-till systems have soil properties that are much more like conventional tillage, and therefore may change pesticide dissipation hardly at all. The discussion has also focused on soil-applied pesticides. A variety of herbicides, insecticides and fungicides are applied post-emergence to weed and crop foliage. There is no apparent reason to assume that dissipation of such foliarly-applied pesticides will be affected by tillage system, except for that part that does reach the surface with spray or subsequent washoff.

REFERENCES

Glotfelty, D. E., A. W. Taylor, B. C. Turner, and W. H. Zoller. 1984. Volatilization of surface-applied pesticides from fallow soil. J. Agr. Food Chem. 32:638-643.

Hamaker, J. W. 1972. Decomposition: Quantitative aspects. p. 253-342. In C.A.I. Goring and J. W. Hamaker (ed.) Organic chemicals in the soil environment. Marcel Dekker, Inc., New York.

Jury, W. A., W. J. Farmer, W. F. Spencer. 1984. Behavior assessment model for trace organics in soil: II. Chemical classification and parameter sensitivity. J. Environ. Qual. 13:567-572.

Kells, J. J., C. E. Rieck, R. L. Blevins, and W. M. Muir. 1980. Atrazine dissipation as affected by surface pH and tillage. Weed Sci. 28:101-104.

Phillips, R. E., and S. H. Phillips (eds.) 1984. No-tillage agriculture. Van Nostrand Reinhold, New York. pp. 301.

Spencer, W. F., and M. M. Claith. 1974. Factors affecting vapor loss of trifluralin from soil. J. Agr. Food Chem. 22:987-991.

Taylor, A. W. 1978. Post-application volatilization of pesticides under field conditions. J. Air Pollut. Control Assoc. 28:922-927.

Turner, B. C., D. E. Glotfelty, A. W. Taylor, and D. R. Watson. 1978. Volatilization of microencapsulated and conventionally applied chlorpropham in the field. Agron. J. 70:933-937.

Willis, G. H., L. L. McDowell, L. A. Harper, L. M. Southwick, and S. Smith. 1983. Seasonal disappearance and volatilization of tozaphene and DDT from a cotton field. J. Environ. Qual. 12:80-85.

CHAPTER 10

EFFECT OF CONSERVATION TILLAGE ON PESTICIDE DISSIPATION

C. S. Helling,
USDA-ARS, Beltsville, Maryland

INTRODUCTION

This is a response paper to D. E. Glotfelty's "The Effects of Conservation Tillage Practices on Pesticide Volatilization and Degradation" (Chapter 9). That report provides a general review of how conservation tillage practices affect soil properties and, in turn, how these changes may affect pesticide dissipation. The impact of tillage management on pesticide uptake by plants was not discussed. This response paper provides some research examples of how tillage has affected soil properties and pesticide persistence. The study of both was not the objective of most investigations, so cause and effect often must be inferred. Major emphasis (and most resource information) is on comparing no-till with conventional moldboard plow tillage since the no-till represents the extreme case of various forms of conservation tillage practices. It should be recognized, however, that imtermediate forms of conservation tillage, e.g., mulch till and ridge till, may more nearly resemble conventional tillage with respect to soil physical and chemical changes.

SOIL PROPERTIES AND PROCESSES

Soil Organic Matter

Conservation tillage requires the maintenance of at least 30% plant residue cover on the soil surface after planting; the

Effects of Conservation Tillage on Groundwater Quality: Nitrates and Pesticides, Terry J. Logan et al., eds.

no-till variation assumes 90–100% residue cover. One major effect of no-till management is, therefore, the development of surface soil with a significantly higher organic matter (OM) content. Table 1 shows the generalized results of three long-term studies illustrating this change. Because plant residues are not mixed into the soil, it is reasonable to expect that the increase in OM content will be most dramatic when the analyzed surface layer is very shallow, e.g., 0–2.5 cm as in Van Doren et al., (1977) results in Table 1. Although OM content near the surface increases in no-till management, that in deeper soil may decrease because crop residues are no longer being incorporated as they are during conventional tillage. It is not clear whether there is a net change in the quantity of soil OM present throughout the Ap horizon depth as a result of tilled versus no-till cropping.

Organic matter analyses are based on the total organic carbon (OC) in soil—both nonliving and living. Soil biomass (the 1–2% of soil OC derived from living microorganisms and plant roots) has been shown (Lynch and Panting, 1980) to be ca. 20–40% greater in soil under direct-drilled grain (i.e., no-till) than when plowed first. These comparisons were made for the 0–5 cm surface soil only.

Table 1. Effect of Tillage on Soil Organic Matter Content in the Surface Horizon.

Elapsed time, y	Soil depth, cm	Relative % OM Conventional	No-till	Reference
5	0–5	X	1.5X	Fleige and Baeumer, 1974
6	0–8	X	1.6X	Slack et al., 1978
10	0–2.5	X	2.0X	Van Doren et al., 1977

Soil pH

Shallow (upper 1-2 cm) soil samples have revealed that pH decreases up to ca. 1 unit after conservation tillage is practiced, especially as a result of prolonged nitrogen fertilization (Blevins et al., 1982; Griffith et al., 1977; Triplett and Van Doren, 1969). Liming apparently readily alleviates this problem.

Soil-Moisture Content and Soil Temperature

The plant residue mulch which overlies the soil surface in non-tilled fields tends to attenuate solar radiation and air currents. The net result is a cooler, wetter soil during the preplant and early growth seasons. Evaporation is reduced, especially in soils of medium texture (sands and clays tend to "self-mulch," i.e., form dry surfaces to restrict evaporation, so the advantage of the plant residue mulch is less for such soils).

Examples of the tendency for high soil-moisture content under no-till management are seen in Figures 1 and 2. In every case over the 6-year period shown in Figure 1, more available moisture was found in no-till plots. In another study (Blevins et al., 1971), shown in Figure 2, higher moisture under no-till management was maintained down to the 90-cm depth and was especially apparent early in the growing season.

Soil temperature is somewhat cooler under no-tillage management, because of mulch shading and also because the higher thermal conductivity of the wetter soil allows heat flow into the subsoil. Thomas, 1985, found that soil temperatures in late June averaged 5°C lower under no-till and maximum soil temperatures dropped from 40° to 32°C.

Bulk Density

The effect of tillage on soil compaction does not seem to be clear cut (Thomas, 1985). With Alfisols and Ultisols, there seems to be little compaction (i.e., an increase in bulk density) when no-tillage is practiced, but some increase may occur in Mollisols. Compaction in Coastal Plain Soils may have reduced the acceptance of no-till in this region (Gebhardt et al., 1985). Gebhardt et al., 1985 noted that although tilled soil is less dense during the early period following plowing, natural subsidence increases the bulk density to a value equal to or greater than that of soil under conservation tillage.

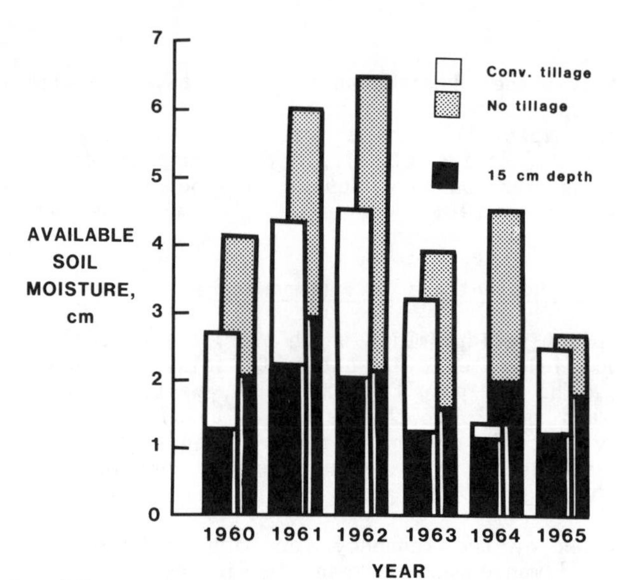

Figure 1. Effect of tillage on available soil moisture at 15 cm (lower histograms) and 46 cm (upper histograms) depths. Average values for June-August in corn plots in Virginia. Adapted from Shear, 1985.

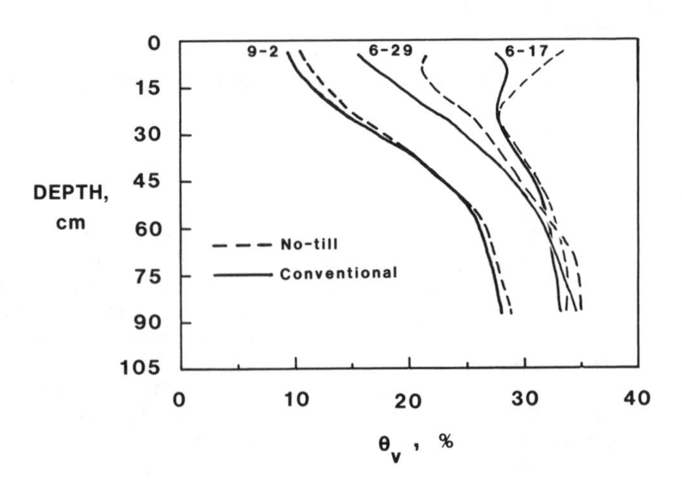

Figure 2. Effect of tillage on volumetric soil-moisture content (θ_v), in Maury silt loam under corn production, for three dates in 1970. Adapted from Blevins et al., 1971.

PESTICIDE DISSIPATION

Loss of pesticides from surface soil can occur by physical transport, degradation, or plant uptake. Transport by leaching or runoff can directly affect water quality, but is considered in other reports. "Persistence," a relative measurement of what proportion of the pesticide remains in soil after a period of time, is determined by how rapidly dissipation occurs. Persistence affects weed control, but excessive persistence may injure the next crop. As an example, atrazine carryover in a wheat-fallow-wheat rotation (where the fallow period is a form of conservation tillage used to replenish soil moisture) would be a major problem for farmers. The two major processes affecting persistence are, in general, volatilization and degradation.

Volatilization

Volatilization is a transport mechanism both within the soil mass and at its surface. Except for such insecticides as carbofuran or carbaryl which may be incorporated with corn during planting, most other pesticides used in no-till culture are surface-applied without further incorporation. Surface dissipation becomes a potentially major loss mechanism for certain pesticides. Microencapsulation, such as with the herbicide alachlor, has reduced this loss and has also prevented some of the entrapment of pesticide within the mulch layer.

Glotfelty (Chapter 9) described some complicating effects of tillage changes (i.e., no-till versus conventional) on pesticide volatilization. Factors increasing the loss potential include the method of pesticide application (surface spray, not incorporated), interception by the mulch layer (less adsorptive than soil, and with more efficient gas exchange) and a moister soil. On the other hand, the slightly cooler temperature, higher OM content (more adsorption), and lower pH (also more adsorption for chemicals such as *s*-triazine herbicides) generally associated with long-term no-till would tend to reduce volatilization. Volatilization seems to have received less research attention than other aspects of pesticide fate in conservation tillage systems. In part this may be due to the more complicated field sampling equipment needed and to some difficulty in accurately estimating total pesticide loss. A research need seems to be careful field monitoring of volatilization loss, particularly as affected by the degree of plant residue cover and by weather variables.

Degradation

Microbial decomposition probably constitutes the major dissipation mechanism for most pesticides used in conjunction with conservation tillage. The changes in soil properties described earlier would be expected to influence microbial ecology and this is what Doran, 1980 found. Analyses of samples from seven U.S. sites led to the results expressed in Table 2. In general, microbial population in the surface 7.5 cm was higher in no-till soils than conventionally tilled soils, as were phosphatase and dehydrogenase enzyme activities, organic C, and potentially mineralizable N. A reversal of these trends occurred in soil from 7.5-15 cm, with definite indication of more anaerobic activity in the subsurface of no-till soils. The high microbial activity in surface no-till soil is consistent with the greater average moisture and OM content, which apparently overcame the generally lower pH and (in the field) soil temperature of no-till soil.

With respect to pesticide biodegradation, most pesticides would probably degrade somewhat faster in the no-till soil than in plowed soil. However, pesticides leaching below ca. 7 cm may degrade more slowly unless anaerobic degradation is an important mechanism.

In some important cases, especially chloro-s-triazine herbicides such as atrazine and simazine, chemical hydrolysis is a significant process of dissipation (Weber and Lowder, 1985). This proceeds faster in acid soils so inactivation (and reduction in weed control) is expected in unlimed no-till fields. Kells et al., (1980) found, when averaging no-till and conventional till soils, that 27% of surface-applied atrazine degraded in soils of pH >6.5, but 35% degraded in soils with pH <5.0, both after 14 days. The formation of nonextractable residue was faster under no-tillage and weed control was apparently less, for the same amount of extractable parent, in the no-till samples. Other investigators (Burnside and Wicks, 1980; Lynch and Panting, 1980) have reported less triazine carryover when using conservation tillage. However, Bauman and Ross, (1983) noted that atrazine persistence was less for

Table 2. Effect of Tillage on Microbial Population at Two Soil Depths (Doran, 1980).

Soil depth, cm	Population ratio, no-till:conventional till					
	Aerobes					
	Total	Aerobic bacteria	Actino-mycetes	Fungi	Facul-tative anaerobes	Denitri-fiers
0-7.5	1.35	1.41	1.14	1.57	1.57	7.31
7.5-15	0.71	0.68	0.98	0.76	1.23	1.77

coulter tillage (as used in no-till) than for chisel or conventional tillage in the first year of application, but after five annual applications, atrazine residues were generally higher in the no-till plots. The explanation for these results is not clear.

Plant Uptake

Plant uptake, through both root uptake and direct vapor phase sorption, of chlorinated insecticides has been demonstrated. Under greenhouse conditions, soybeans only removed 0.02-0.61% of the insecticide from soil (Beall and Nash, 1971). Uptake of carbofuran by corn averaged only ca. 0.14% of that in soil, and dieldrin uptake was approximately 0.01% (Caro et al., 1976). Uptake of pesticides was reportedly greater as the chemical's polarity decreases (Briggs et al., 1982); this seems to contradict the carbofuran (more polar)--dieldrin (less polar) observations of Caro et al., (1976). Perhaps the distinction lies in root uptake, which may resemble pesticide adsorption onto soil organic matter, and translocation to the leaves, which is probably more important for polar chemicals. For ionizable compounds such as the herbicide chlorsulfuron, soil pH can significantly affect both uptake into crops and degradation in soils (both being greater for chlorsulfuron at pH 5.6-5.9 than at pH 7.5) (Frederickson and Shea, 1986). Some of the complexities of relating uptake of chemicals as measured in laboratory vs. field conditions have been reviewed (Kloskowski et al., 1981). It seems unlikely that the quantity of pesticides removed via plant uptake will constitute a major loss pathway, but this issue is worth research attention.

SUMMARY

Pesticide dissipation changes that are caused by conservation tillage practices are of uncertain magnitude. Among reasons for this uncertainty are the wide range of tillage practices which, in turn, lead to differing impacts on soil physical and chemical properties. The extreme situation, no-till farming, can lead to higher OM content, cooler soil temperature, higher moisture content, lower pH, and greater microbial activity; these effects are most pronounced nearest the surface. Pesticide degradation is probably somewhat faster in no-till than in conventional tillage.

REFERENCES

Bauman, T. T., and M. A. Ross. 1983. Effect of three tillage systems on the persistence of atrazine. Weed Sci. 31:423–426.

Beall, M. L., Jr., and R. G. Nash. 1971. Organochlorine insecticide residues in soybean plant tops: root vs. vapor sorption. Agron. J. 68:460–464.

Blevins, R. L., D. Cook, S. H. Phillips, and R. E. Phillips. 1971. Influence of no-tillage on soil moisture. Agron. J. 63:593–596.

Blevins, R. L., G. W. Thomas, M. S. Smith, W. W. Frye, and P. L. Cornelius. 1982. Changes in soil properties after 10 year non-tilled and conventionally tilled corn. Soil Tillage Res. 2.

Briggs, G. G., R. H. Bromilow, and A. A. Evans. 1982. Relationships between lipophilicity and root uptake and translocation of nonionized chemicals by barley. Pestic. Sci. 13:497–504.

Burnside, O. C., and G. A. Wicks. 1980. Atrazine carryover in a reduced tillage crop production system. Weed Sci. 28:661–666.

Caro, J. H., A. W. Taylor, and H. P. Freeman. 1976. Comparative behavior of dieldrin and carbofuran in the field. Arch. Contamin. Toxicol. 3:437–447.

Doran, G. W. 1980. Soil microbial and biochemical changes associated with reduced tillage. Soil Sci. Soc. Am. J. 44:765–771.

Fleige, H., and K. Baeumer. 1974. Effect of no-tillage on organic carbon and total nitrogen content, and their distribution in different N fractions in loessial soils. Agr. Ecosystems 1:19–29.

Frederickson, D. R., and P. J. Shea. 1986. Effect of soil pH on degradation, movement, and plant uptake of chlorsulfuron. Weed Sci. 34:328–332.

Gebhardt, M. R., T. C. Daniel, E. E. Schweizer, and R. R. Allmaras. 1985. Conservation tillage. Science 230:625–630.

Griffith, D. R., J. V. Mannering, and W. C. Moldenhauer. 1977. Conservation tillage in the eastern corn belt. J. Soil Water Conserv. 32:321–326.

Kells, J. J., C. E. Rieck, R. L. Blevins, and W. M. Muir. 1980. Atrazine dissipation as affected by surface pH and tillage. Weed Sci. 28:101–104.

Kloskowski, R., I. Scheunert, W. Klein, and F. Korte. 1981. Laboratory screening of distribution, conversion and mineralization of chemicals in the soil-plant-system and comparison to outdoor experimental data. Chemosphere 10:1089-1100.

Lynch, J. M., and L. M. Panting. 1980. Cultivation and the soil biomass. Soil Biol. Biochem. 12:29-33.

Shear, G. M. 1985. Introduction and history of limited tillage. p. 1-14. In A. F. Wiese (ed.), Weed control in limited-tillage systems, Monogr. No. 2, Weed Sci. Soc. Am., Champaign, Ill.

Slack, C. H., R. L. Blevins, and C. E. Rieck. 1978. Effects of soil pH and tillage on persistence of simazine. Weed Sci. 26:145-148.

Thomas, G. W. 1985. Managing minimum-tillage fields, fertility, and soil type. p. 211-226. In A. F. Wiese (ed.), Weed control in limited-tillage systems, Monogr. No. 2, Weed Sci. Soc. Am., Champaign, Ill.

Triplett, G. B., Jr., and D. M. Van Doren, Jr. 1969. Nitrogen, phosphorus and potassium fertilization of non-tilled maize. Agron. J. 61:637:639.

Van Doren, D. M., G. B. Triplett, Jr., and Jr. E. Henry. 1977. Influence of long term tillage and crop rotation combination on crop yields and selected soil parameters for an Aeric Orchraqualf soil. Ohio Agric. Res. Dev. Cent. Res. Bull. 1091.

Weber, J. B., and S. W. Lowder. 1985. Soil factors affecting herbicide behavior in reduced-tillage systems. p. 227-241. In A. F. Wiese (ed.), Weed control in limited-tillage systems, Monogr. No. 2, Weed Sci. Soc. Am., Champaign, Ill.

CHAPTER 11

PROCESSES INFLUENCING PESTICIDE LOSS WITH WATER UNDER CONSERVATION TILLAGE

R. J. Wagenet,
Cornell University, Ithaca, New York

INTRODUCTION

Adoption of conservation tillage practices usually infers increased reliance upon, although not necessarily increased usage of, pesticides, particularly the application of herbicides for weed control. The environmental fate of these chemicals is relatively unstudied under the particular physical, chemical and biological processes operative in conservation tillage (CT) systems, making it necessary to use information developed under conventional agricultural practices to infer appropriate management and regulatory decisions. Although some of this information is quite applicable to CT systems, several rather major issues remain unresolved, and merit increased research attention if the environmental impact of pesticide usage in CT systems is to be minimized.

The Federal Water Pollution Control Act Amendments passed by Congress in 1972 (USEPA, 1972) required states to identify non-point sources of pollution and develop plans to control such sources (Sec. 208). Among the identified possible pollutants are a wide range of agricultural chemicals, including pesticides. As a result, substantial effort has been expended since the early 1970's measuring pesticide emissions from agricultural fields, with almost all this effort focused on conventional tillage systems. It has only been during the last 3-5 years, during which time CT has been more widely promoted and accepted that a few studies have attempted to quanitfy pesticide losses by water under this increasingly widespread practice. It is clear that the unique nature of such systems,

Effects of Conservation Tillage on Groundwater Quality: Nitrates and Pesticides, Terry J. Logan et al., eds.

including heavier pesticide applications, reduced mechanical manipulation of soil and crop residue, and altered water retention and flow properties will result in pesticide behavior that cannot necessarily be inferred from studies accomplished in conventional systems. Yet the basic physical, chemical and biological processes that must be involved are well documented, providing us with an intellectual framework within which we can organize both knowns and unknowns regarding pesticide fate under CT.

PESTICIDE USAGE UNDER CONSERVATION TILLAGE

A number of herbicides commonly used under conventional tillage are also used under CT, with a representative and not necessarily comprehensive list presented in Table 1. Notably absent from the list are those chemicals, such as EPTC and trifluralin, that must be soil incorporated to be effective against their targets. The accumulation of crop residues and the desire to minimize mechnaical manipulation of the soil generally precludes the use of such compounds under reduced tillage situations. There is also little specific information available on the use of both insecticides and fungicides in CT systems, although it appears that slightly increased insecticide use may be necessary in response to such pests as cutworm and armyworm. It appears, though, that weeds and herbicides have been the first concern of both the grower and the researcher, and the application, management and fate of other pesticides in CT systems is not well studied.

Summary statistics of the quantities of these chemicals used in CT programs are not easily available, and the list in Table 1 is certainly incomplete given the very crop-specific and pest-specific issues associated with chemical application programs. A factor important to pesticide loss by water is that both soil and foliar applied chemicals of a range of water solubilities are used. The implications of this will be discussed below.

Field studies of pesticide fate have only been accomplished under a very limited number of environmental conditions for most of the chemicals listed in Table 1. Very few of these studies have been under CT. The behavior of most pesticides, whether used in CT or conventional argicultural systems, are relatively unstudied for more than a few of the field conditions in which the pesticide will be used.

PROCESSES AFFECTING PESTICIDE FATE UNDER CONSERVATION TILLAGE

Movement beyond the point of application of the pesticides used in conservation tillage systems is determined by water losses from the field. Water can exit the field either as surface runoff, carrying pesticide dissolved in the water or sorbed to particulate soil material, or water can drain

Table 1. Representative Herbicides Used in Conservation Tillage (CT) Systems.

Common Name	Trade Name[1]	Water solu. ppmw	Applied to: formulation
Alachlor	LASSO	242	Soil; aq. sol. or dry granule
Atrazine	ATTREX	33	Soil; aq. solution
Cyanazine	BLADEX	171	Soil or foliar; aq. solution
2,4-D (acid)	-	900	Soil; aq. or oil solution
2,4-D (salt)	WEEDAR 64	sol[3]	Soil; aq. or oil solution
2,4_d (ester)	ESTERON 99	10-500[2]	Soil; aq. or oil solution
Dicamba (salt)	DANVEL	sol[3]	Foliar spray; aq. solution
Diuron	KARMEX	42	Soil; aq. sol. or dry granule
Glyphosate	ROUNDUP	12,000	Foliar; aq. solution
Hexazinone	VELPAR	33,000	Foliar; aq. solution
Linuron	LOROX	75	Soil; aq. solution
Metolachlor	DUAL	530	Soil; aq. solution
Metribuzin	SENCOR	1,220	Soil; aq. solution
Molinate	ORDRAM	800	Soil; emulsifiable conc.
Paraquat	ORTHO PARAQUAT	sol[3]	Foliar; aq. solution
Picloram (salt)	TORDON	sol[3]	Foil or foliar; dry granules or aq. solutions and sprays
Propachlor	RAMROD	580	Soil; granular or wettable powder
Propham	CHEM HOE	250	Soil; aq. spray or granular
Simazine	PRINCEP	5	Soil; aq. sol. or dry granules
Terbutryn	IGRAN	58	Soil or foliar; aq. solution

1. Example of one name only
2. Assumed from propetie of similar cmpds.
3. sol. means solubility > 10%wt/vol.

vertically through the soil profile, transporting primarily dissolved chemical to deeper soil depths and eventually, if the chemical persists long enough, to ground water.

While the partitioning of water between drainage and surface runoff, and the amount of eroded material carried in the runoff are important, pesticide loss by water is also greatly dependent upon factors and processes that vary according to chemical type, soil type, and crop residue. These influence not only the water flow regime, but also the mass of chemical available to be transported by water. There have been a number of studies in conventional tillage systems focusing on the integrated effects of these factors with respect to pesticide loss via surface runoff (Wauchope, 1978), and very few studies on leaching losses of the chemicals.

Several factors influence the mass of pesticide susceptible to loss via surface runoff or leaching. First, not all the chemical is instantaneously available for movement, with both the chemical's mode of application (i.e., wettable powder, aqueous solution) and its solubility being important factors. Second, mediating processes such as pesticide persistence (degradation) in the soil-water system, interaction of pesticide with soil mineral and organic particles (sorption) and plant uptake of pesticides all influence the amount of chemical available for transport. Volatilization, degradation and plant uptake are discussed elsewhere in this workshop, with attention here focused on sorption and transport processes.

PESTICIDE LOSSES VIA SURFACE RUNOFF

Most pesticides interact quite strongly with soil and sediment surfaces, resulting in the partitioning of pesticide mass between aqueous and sorbed phases (Fig. 1). While concentrations of pesticide in runoff water may be in the order of ug/ml, concentrations in the sorbed phase (sediment) of the runoff can be orders of magnitude larger, of the order of thousands of ug/g soil. However, sediment is usually a very small fraction of the total runoff weight and therefore contributes little total mass to the amount of chemical lost during the runoff process. This indicates that, with the exception of highly insoluble or very strongly sorbed chemicals, soil, water, and crop management practices that reduce soil losses, but not runoff volumes, may have little effect on pesticide losses. However, most land management practices that reduce erosion also reduce runoff volume, although not proportionally (Stewart, 1975). It is fortunate that pesticides such as trifluralin, endrin and toxaphene which are toxic to aquatic life are also extremely insoluble pesticides which are controlled by erosion control measures.

No-till cropping drastically reduces both erosion and runoff, with concomittant near elimination of pesticide losses (Foy and Hiranpradit, 1977; Triplett et al, 1978; Edwards, 1972. Minimum tillage systems sufer some surface losses of sediment

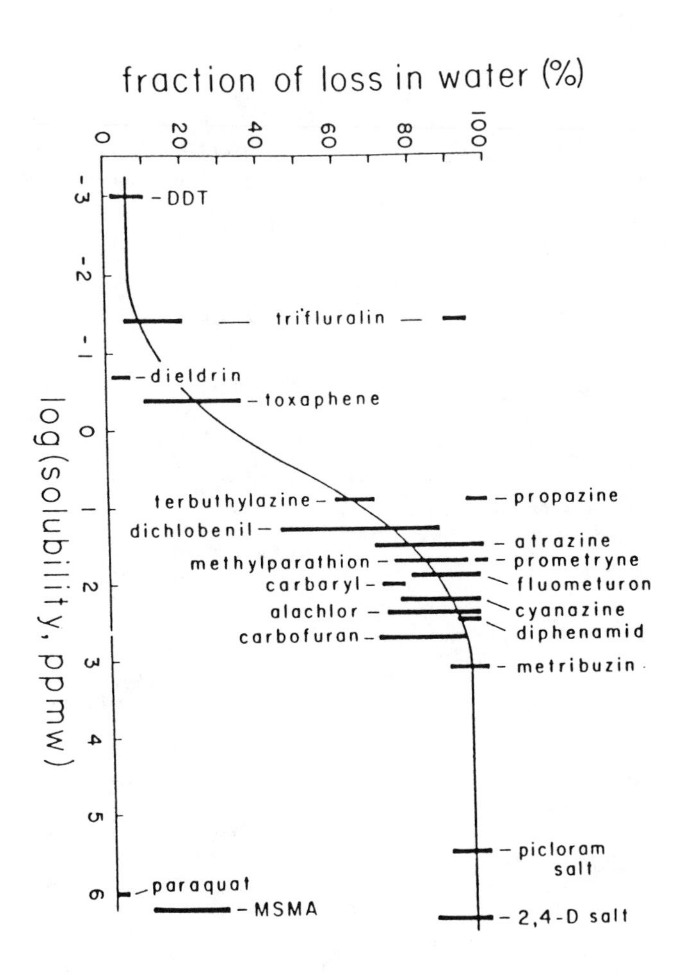

Figure 1. Partitioning of pesticides between sediment and water in runoff samples, with the range of reported literature values indicated by the soild bars (from Wauchope et al., 1985).

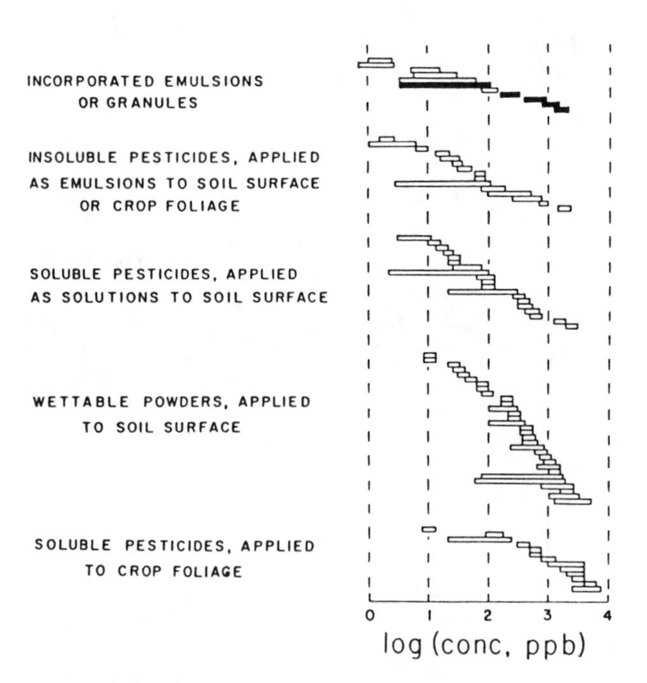

Figure 2. Effect of application formulation on pesticide loss in runoff events. Each bar is a different pesticide, and the range is for different sites, rates and other conditions. Narrow ranges could represent a smaller number of experiments. Dark bars are for incorporated granular pesticides (from Wauchope et al., 1985).

water and pesticide (Smith et al., 1978; Ritter et al., 1974), but the magnitude of these losses is not yet well quantified for a range of cases. Clearly, though the amount of soil covered by crop residues play a major role in determining the effect of management practices on runoff volume and quality (Baker et al., 1978). The residue not only influences soil loss, it also apparently provides a source of pesticide that is easily mobilized by applied water (Martin et al., 1978; Bovey et al., 1974). The implications of such interactions will be discussed shortly.

A very comprehensive review by Wauchope (1978) of studies related to pesticide loss by surface runoff from conventionally tilled agricultural fields contains a number of observations relevant to CT. First, generalized categories of potential pesticide loss by surface runoff were developed as a function of the pesticide application formulation (Fig. 2). These general categories may also be appropriate for CT systems, particularly for those pesticides that are applied as wettable powders or that are relatively insoluble in water (applied as emulsions). The following hierarchy was proposed: (1) Pesticides formulated as wettable powders (all are herbicides) consistently show the highest long-term losses of any general class of herbicides, with losses of up to 5% of applied chemical on moderately sloping fields (10-15%), and up to 2% on fields of 3% slope or less (an observation partly confirmed for atrazine and simazine in systems by Triplett et al. (1978); (2) pesticides usually applied as emulsions (most are foliar applied inscticides) show the next worst long-term surface loss, of 1% or less; (3) Water soluble pesticides applied as aqueous solution usually show surface runoff losses of 0.5% or less, with the exception of ionic compounds, like paraquat, that strongly sorb to soil mineral or organic surfaces.

Wauchope also identified that single storm losses were particularly important to pesticide movement to surface waters, particularly when these storms were within two weeks of pesticide application, with at least 1 cm of rain and a runoff volume of 50% or more of the precipitation. Whether such relationships will hold absolutely for CT systems is unclear, particularly recognizing the influence that added organic matter and crop debris will have in increasing infiltration rates and decreasing surface runoff of water. However, if any surface runoff of water occurs, it is a logical, although presently only qualitatively useful, assumption that applied pesticides will follow these categories of surface runoff susceptibility. Quantitative estimation of pesticide in runoff will require further indentification of pesticide/organic matter interactions, enhanced degradation, and leaching losses (as described below).

The importance of added organic matter with respect to pesticide loss from CT systems is not yet clearly defined, and several possible influences can be postulated. All studies of pesticide loss by surface runoff from conventionally tilled agricultural fields must be applied to CT systems with due consideration for these influences. First, the sorption of pesticides has been shown to increase as the organic matter content of a soil increases, but with the exception of a few studies in peat soil, this generalization is limited to studies accomplished on soils with relatively low organic matter contents (<5%). According to classic sorption isotherm characterizations of this process, the net effect of such interaction is to retard the leaching of the chemical and prolong its presence at the soil surface, thereby increasing the opportunity for loss by surface runoff. If this process holds true in nature, it can be anticipated that increased quantities of decomposing crop resudues will enchance the opportunity for pesticide-organic matter interaction, and will result in more chemical being retained in shallow soil depths. This may well reduce the opportunity for chemical leaching, but increase the opportunity for pesticide to be lost in surface runoff, either as material sorbed to eroded sediment, or dissolved into surface runoff water. The relative importance of these effects will vary according to the chemical type, the crop residue, and the degree of residue decomposition. Such factors are nearly unstudied for the chemicals used in CT.

The efficacy of the applied herbicide can be decreased as the quantity of organic debris at the soil surface increases (Banks and Robinson, 1984). This is the result of reduced quantities of applied chemical reaching the soil surface, and is particularly important for materials absorbed by the plant roots. Such effects can result in recommendations that herbicide application rates be increased, with there being little documentation of the environmental fate of the extra chemical. Additionally, the interception by the crop residue of applied pesticides establishes an increased opportunity for these chemicals to be lost in surface runoff rather than enter the soil. Reformulation of existing herbicides to alter their application mode or promote their slow release, or the development of new, highly selective, postemergence compounds that will be precisely applied at very low rates (Burnside, 1984) may help reduce losses by surface runoff. An issue often neglected is the effect that the mulch itself has on weed populations, and the resulting need for herbicide, which may be less than under conventional tillage systems.

There exists an additional complicating factor of potentially quite significant importance. Preliminary findings by Jury and co-workers at UC-Riverside (Jury, 1986) indicate that some pesticides may complex with the soluble organic fraction of soil humus (in a manner analagous to metal chelation), and actually be transported deeper within the soil profile than would be expected from classical interpretation of the pesticide sorption process. Such an effect would be the opposite of the commonly expected influence of organic matter, and could result in added deep leaching of certain pesticides. Again, such processes are unstudied for either the chemicals or the organic matter present in CT systems.

A second major influence of the increased organic matter content will be the effect upon soil microbiological populations, and the degradation rate of pesticides used as microbial substrates. The mass of chemical available for overland transport or leaching may be greatly reduced as populations of decomposer organisms respond to increased organic matter levels. It is also unclear whether the expected increased sorption discussed above will "protect" pesticides from degradation by removing it from the solution to the sorbed phase. Such factors have not been generally studied because of the relatively low organic matter content in most agricultural soils, but are worthy of increased attention as CT is more widely used, and will certainly influence the mass of pesticide lost by water.

PESTICIDE LOSSES VIA LEACHING

It is often true that a consequence of CT practices is an increase in infiltration rates, a decrease in soil evaporation and surface runoff, and a resulting increase in the quantity of water in the soil profile. It can be expected that the drainage of water beyond the root zone will often also increase in such systems, although this may not always be the case depending upon the other components of the water budget. Clearly though, as more water is introduced to the soil profile, the opportunity for leaching of applied chemicals will increase.

Pesticide loss via leaching has been very infrequently studied, even in conventional tillage systems. Whether it represents a significant avenue of pesticide loss in CT systems is not clear. It is true that some pesticides, such as paraquat, atrazine and metolachlor, are relatively non-mobile in soils of average organic matter content, and can be expected to be even less mobile in soils of higher organic matter content (assuming the absence of a chelation-type mechanism as discussed above). Other pesticides are susceptible to leaching under a wide range of conditions. As with issues related to surface runoff, better identification of the nature of pesticide sorption in soils of high organic matter will be crucial information needed to assess potential pesticide mobility under CT.

A number of other issues also influence the degree of leaching loss of pesticides from CT systems. One major concern focuses upon the establishment in these systems of macropores, relatively large and vertically continuous soil pores that result from, for example, earthworm activity. Such macroporosity is normally disrupted under conventional tillage by the mechanical mixing of the upper soil profile during normal plowing or discing operations. When left undisturbed, rather permanent channels, or macropores, will certainly be established in the root zone and will probably be continuous from the soil surface through the first meter of the profile. These macropores can influence pesticide leaching in at least two ways.

Under one possible scenario, when high intensity rainfall events occur, and water application rates exceed soil infiltration rates, it has been hypothesized that the macropores will conduct water and any dissolved chemical downward to deeper soil depths. This "short-circuiting" or "by-pass" flow could result in transport of surface-applied chemicals to soil depths generally less biologically active than the surface (0-30 cm). The opportunity for microbiological degradation is commensurately decreased, and sorption of the pesticide, greatest where organic matter is greatest, is also less a factor in retarding the chemical's movement. The net result of this combination of events would be leaching of a greater pesticide mass at a more rapid rate than would normally be expected. However, a different effect of macropore flow can also be postulated. If a soluble pesticide is applied to the soil surface, and not immediately leached by water, the chemical will dissolve into solution and, through diffusion, become distributed between large and small pores in the surface soil depth. Any application of water that then results in macropore flow will result in little displacement of pesticide residing in the smaller pores, while water with relatively low pesticide content will move readily to a deeper depth. In such cases, the macropore actually will result in less leaching of applied pesticide than would result from "classical" miscible displacement processes of transport through the soil matrix. The establishment, prevalence and influence of macropores upon pesticide movement in CT systems has not been adequately studied, but deserves further conditions of water and pesticide at the soil surface under CT programs.

Even in the absence of a macropore mechanism of leaching, increased water flow through soil could result in increased losses of applied chemical. The absolute magnitude of these losses is difficult to estimate. A Canadian study (Frank et al., 1982) of pesticide content of waters derived primarily from conventionally tilled agricultural watersheds found that, of pesticides lost to water, 60% derived from storm runoff waters, 18% from internal drainage (leaching) and the remainder from spills and associated careless handling practices. It has additionally been reported (Wauchope, 1978) that pesticide loss from surface runoff is approximately 2-5% of the total amount applied, leaving only a small total mass of applied chemical to be potentially lost in the drainage. The balance (95-98% of that applied) presumably is degraded, taken up by the plant, volatilized, or remains as a residue. Yet, these percentages appear very tenuous, particularly considering observed spatial variability (summarized by Rao and Wagenet, 1985) of leaching and other processes that determine pesticide fate in field soils. Very few field studies have been accomplished that can be used to confirm or deny such values. It is quite poosible that particular circumstances of permeable soils, high pesticide application rate, and increased water infiltration could result in leaching losses of pesticide far greater than those reported to date. Field studies that involve direct measurement of leaching, and which employ enough samples to quantify spatial variability (Rao and Wagenet, 1985), will be required to resolve this question.

MODELING PESTICIDE LOSS BY WATER

It is not possible to conduct experiments under every conceivable set of environmental conditions in which pesticides may be applied, and this includes CT. Models of pesticide fate have been proposed as substitutes for experiments, and as tools in the regulatory process.

Mathematical models of surface runoff and leaching of pesticides have been constructed, tested and used with varying degrees of success. The formulation of each model varies according to the objectives of the modeling exercise and the professional training and biases of the model developer. The result has been a collection of approaches applicable to description of surface runoff processes, and a second body of efforts focusing on leaching processes. Comprehensive simultaneous descriptions have not often been attempted, and when they have, have resulted in complex, data-intensive models not easily used by other than the developer.

A number of models of processes affecting surface runoff and resultant pesticide loading of surface waters have been proposed (Bruce et al., 1975; Frere et al., 1975; Donigian and Crawford, 1976; Adams and Kurisu, 1976; Donigian et al., 1977; Leonard and Wauchope, 1980; Wauchope and Leonard, 1980; Haith, 1980; Haith, 1986). In almost all cases, the models represent a compromise between the available data, which is often quite sparse and variable, and the need for a predictive tool that can be employed across different soils, climates and pesticides. Mixed results have been obtained, and to date there is apprently no increased predictive capability obtained by using models that are more mechanistic and data intensive versus less mechanistic, empirical or regression methods. This is partially an indictment of our understanding of field scale processes related to pesticide loss, as well as indication of our limits in description of surface hydrological processes.

Soil leaching models of pesticide fate suffer similar problems, although the basic physical, chemical and biological processes in the soil unsaturated zone are perhaps better defined than for surface hydrology. Models useful on a field scale exist in both mechanistic (Carsel et al., 1984; Wagenet and Hutson, 1986) and non-mechanistic (Rao et al., 1976; Nofziger and Hornsby, 1986) forms, although care must be utilized in the situations to which these models are applied. All are relatively untested under field conditions, and the non-mechanistic versions are generally intended for qualitative educational purposes rather than quantitative regulatory purposes. A number of solute transport models have been proposed (reviewed by Addiscott and Wagenet, 1985) that are intermediate to these two extremes, but have yet to be applied to pesticide leaching by water. The spatial variability of soil processes also has generated interest in stochastic or probabilistic approaches for description of chemcial leaching in soil (e.g., Jury et al., 1985) or surface loss of pesticide (Mills and Leonard, 1984). These approaches may prove to be the most useful, as they show promise as descriptors of spatially variable processes, yet are neither as mathematically cumbersome or computationally demanding as current mechanistic models. They are not yet formulated in a manner useful in description of pesticides under CT conditions, but deserve attention as models of pesticide loss in CT systems are developed.

There are a number of major issues presently not well considered in modeling exercises of pesticide loss by water. First, the management of the soil surface, including influences of organic debris on water, heat, and pesticide fluxes across the soil-air boundary, is not very well modeled. This issue is particularly crucial as models of CT systems are developed. Second, there is incomplete consideration in most models of the spatial variation at any one site of such variables as pesticide application pattern and uniformity, soil leaching processes, degradation, sorption and other mediating factors. As process level models are formulated that include description of these basic processes, it will be important to identify, for example, the expected mean and variance of these processes, perhaps as a function of soil type, climate and pesticide form. Efforts to synthesize and summarize data with such issues in mind have been accomplished (e.g., Leonard et al., 1979) or are underway. If research results demonstrate an inability to formulate accurate statistical descriptions of field scale processes, it may be necessary to go formulate models in less-mechanistic forms, such as proposed by Wauchope and Leonard (1980) and Haith (1986). Finally, there is a substantial gap between our need to understand field scale processes of pesticide losses by water, and the economic realities of field sampling programs required to satisfy those needs. Resources do not presently exist to conduct the number of field studies, with all the different chemicals that are possible, at the intensity required to fully test and verify the predictive capabilities of any model. It appears that our ability to intelligently and effectively use models to predict pesticide loss from conservation (or any other) tillage system will always be somewhat constrained by this factor, and redefinition of model output, in terms of statistical uncertainty of the output, may be a necessary step as models are used in risk assessments or for other regulatory purposes.

RESEARCH NEEDS

The most immediate research needs related to pesticide loss by water under CT may be summarized in the following manner:

1. The influence of organic debris at the soil surface upon pesticide partitioning between aqueous and sorbed phases needs to be better quantified. The effects of this partitioning on the proportion of applied chemical lost by surface runoff and leaching, and any associated changes in efficacy of the chemical, require further study.

2. The establishment, prevalence and influence of macropores upon pesticide leaching in CT systems needs to be quantified.

3. Field studies that involve direct measurement of leaching, and which are designed to consider the spatial variability of the soil-water-pesticide system, are needed to quantify leaching losses of pesticide under a range of soil, climate, crop and chemical conditions.

4. Improved models useful in predicting pesticide loss by surface runoff and leaching are needed. Application of these models to CT systems will require that particular attention be paid to the influence upon pesticide fate of organic debris at the soil surface, and the variability of soil processes.

REFERENCES

Adams, R. T. and F. M. Kurisu. 1976. Simulation of pesticide movement on small agricultural watersheds. Environ. Systems Lab., Sunnyvale, CA. Prepared for USEPA, Athens, GA. Pub. No. EPA-600/3-76-066.

Addiscott, T. M. and R. J. Wagenet. 1985. Concepts of solute leaching in soils: A review of modeling approaches. J. Soil Sci. 36:411-424.

Baker, J. L., J. M. Laflen and H. P. Johnson. 1978. Effect of tillage systems on runoff losses of pesticides, a rainfall simulation study. Trans Am. Soc. Agric. Eng. 21:886-892.

Banks, P. A. and E. L. Robinson. 1984. The fate of oryzalin applied to straw-mulched and nonmulched soils. Weed Sci. 32:269-272.

Bovey, R. W., E. Burnett, C. Richardson, M. G. Merkle, J. R. Baur and W. G. Knisel. 1974. Occurrence of 2,4,5-T and picloram in surface runoff of the Blacklands of Texas. J. Environ. Qual. 3:61-64.

Bruce, R. R., L. A. Harper, R. A. Leonard, W. M. Snyder and A. W. Thomas. 1975. A model for runoff of pesticides from small upland watersheds. J. Environ. Qual. 4:541-548.

Burnside, O. C. 1984. Requirements for agricultural chemicals in changing agricultural production systems - Current state of art. pp. 49-71. In: G. W. Irving (eds.). Changing agricultural production systems and the fate of agricultural chemicals. Agric. Res. Inst., Chevy Chase, MD. pp. 160.

Carsel, R. F., D. N. Smith and M. N. Lorber. 1984. User's manual for the Pesticide Root Zone Model (PRZM) Release 1. USEPA, Athens, GA. Pub. No. EPA-600/3-84-109.

Donigian, A. S., Jr., D. C. Berzerlein, H. H. Davis, Jr., and N. H. Crawford. 1977. Agricultural runoff management (ARM) model - Version II. Refinement and testing. Hydrocomp, Inc., Palo Alto, Calif. Prepared for USEPA, Athens, GA. Pub. No. EPA-600/3-77-098.

Donigian, A. S., Jr. and N. H. Crawford. 1976. Modeling pesticide and nutrients on agricultural lands. Hydrocomp, Inc. Palo Alto, CA. Prepared for USEPA, Athens, GA. Pub. No. EPA-6002-76-043.

Edwards, W. M. 1972. Agricultural chemical pollution as affected by reduced tillage systems. pp. 30-40. In: Proc. No-Tillage Systems Symp., Feb. 21-22, 1972, Ohio State University.

Foy, C. L. and H. Hiranpradit. 1977. Herbicide movement with water and effects of contaminant levels on nontarget organisms. Nat. Tech. Inf. Serv. PB-263-285, Jan.

Frank, R., H. E. Braun, M. Van Hove Holdrinet, G. J. Sirons, and B. D. Ripley. 1982. Agriculture and water quality in the Canadian Great Lakes basin: V. Pesticide use in 11 agricultural watersheds and presence in stream water, 1975-1977. J. Environ. Qual. 11:497-505.

Frere, M. H. 1978. Models for predicting water pollution from agricultural watersheds. pp. 501-509. In: Conf. on Modeling and Simulation of Land, Air and Water Resources Systems. Int. Fed. Inf. Process. Ghent, Belgium.

Haith, D. A. 1980. A mathematical model for estimating pesticide losses in runoff. J. Environ. Qual. 9:428-433.

Haith, D. A. 1986. Simulated regional variations in pesticide runoff. J. Environ. Qual. 15:5-8.

Jury, W. A. 1986. Personal communication.

Jury, W. A., G. Sposito and R. E. White. 1986. A transfer function model of solute transport through soil. 1. Fundamental concepts. Water Resour. Res. 22:243-247.

Leonard, R. A. and W. D. Wauchope. 1980. The pesticide submodel. pp. 88-112. In: W. G. Knisel (eds.). CREAMS: A field-scale model for chemicals, runoff and erosion from agricultural management systems. U.S.D.A. Conservation Res. Rep. No. 26, 640 pp.

Leonard, R. A., G. W. Langdale and W. G. Fleming. 1979. Herbicide runoff from Upland Piedmont watersheds - Data and implications for modeling pesticide transport. J. Environ. Qual. 8:223-229.

Martin, C. D., J. L. Baker, D. C. Erbach and H. P. Johnson. 1978. Washoff of herbicides applied to corn residue. Trans. Am. Soc. Agric. Eng. 21:1164-1168.

Mills, W. C. and R. A. Leonard. 1984. Pesticide pollution probabilities. Trans. Amer. Soc. Agr. Eng. 27:1704-1710.

Nofziger, D. L. and A. G. Hornsby. 1986. A microcomputer-based management tool for chemical movement in soil. Appl. Agric. Res. 1:50-56.

Rao, P. S. C., J. M. Davidson and L. C. Hammond. 1976. Estimation of nonreactive and reactive solute front locations in soils. pp. 235-241. In: Proc. Hazard Wastes Res. Symp., Tucson, AZ. Pub. No. EPA-600/19-76-015.

Rao, P. S. C. and R. J. Wagenet. 1985. Spatial variability of pesticides in field soils: Methods for data analysis and consequences. Weed Sci. 33 (Supp. 2): 18-24.

Ritter, W. F., H. P. Johnson, W. G. Lovely and M. Molnau. 1974. Atrazine, propachlor, and diazinon residues on small agricultural watersheds. Environ. Sci. Technol. 8:38-42.

Smith, C. N., R. A. Leonard, G. W. Langdale and G. W. Bailey. 1978. Transport of agricultural chemicals from small upland Piedmont watersheds. USEPA. EPA-600/3-78-056. 363 pp.

Stewart, B. A. 1975. Control of water pollution from cropland. Vol. I. A manual for guideline development. U.S.D.A. Rep. ARS-H-5-1. pp. 111.

Triplett, G. B., B. J. Conner and W. M. Edwards. 1978. Transport of atrazine and simazine in runoff from conventional and no-tillage corn. J. Environ. Qual. 7:77-84.

U.S. Environmental Protection Agency. 1972. Federal Water Pollution Control Act Amendments PL 92-500. Oct. 18, 1972.

Wagenet, R. J. and J. L. Hutson. 1986. Predicting the fate of non-volatile pesticides in the unsaturated zone. J. Environ. Qual. 15:315-322.

Wauchope, R. D. 1978. The pesticide content of surface water draining from agricultural fields - A review. J. Environ. Qual. 7:459-472.

Wauchope, R. D. and R. A. Leonard. 1980. Maximum pesticide concentrations in agricultural runoff: A semi-empirical prediction formula. J. Environ. Qual. 9:665-672.

Wauchope, R. D., L. L. McDowell and L. J. Hagen. 1985. Environmental effects of limited tillage. pp. 266-281. In: A. F. Wiese (ed.) Weed Control in Limited - Tillage Systems. Weed Sci. Soc. Am. Monograph Series Number 2. pp. 297.

CHAPTER 12
EFFECTS OF CONSERVATION TILLAGE ON PESTICIDE LOSS WITH WATER

R. D. Wauchope,
USDA, Southeast Watershed Research Center, Tifton, Georgia

INTRODUCTION

Wagenet (Chapter 11) has given a useful overview of the effect of conservation-tillage (CT) practices on pesticide losses in water. His paper emphasizes the need for more studies comparing losses between tillage systems, a more detailed look at crop residue effects, and the macropore phenomemon. He also discusses the question of how to represent (describe) spacial variability, an issue that is at the heart of current modeling efforts.

Rather than "critique" Wagenet's paper *per se*, I prefer to add emphasis and detail in those areas he has covered which seem to have high priority research needs.

DEFINING THE ISSUE

We are concerned with the question,

> Does the adoption of conservation tillage practices lead to increased nonpoint inputs of pesticides, especially herbicides, to surface water and groundwater?

For the purposes of this discussion, "conservation tillage" (CT) includes "reduced tillage" (RT) and "no-till" (NT), and RT includes any practice which leaves at least 30% crop residue

Effects of Conservation Tillage on Groundwater Quality: Nitrates and Pesticides, Terry J. Logan et al., eds.

(Agrichemical Age, 1984; Allmaras and Dowdy, 1985; Crosson, 1981).

Among all the other concerns and issues related to CT adoption, we want to minimize nonpoint pesticide loads to surface and groundwaters. This is not necessarily because we know the environmental impacts of such an increase will be negative. In many cases we simply do not know what, if any, the effects are (Wauchope, 1978; Wauchope et al., 1985). In general, it appears that current loadings of pesticides to surface waters in the US are not a serious problem. Thus, if CT does not increase the loadings significantly, this aspect of CT will presumably not be a problem.

The current concern over pesticides in groundwater complicates things, however. We can say with some confidence that we know something about surface-water (runoff) pesticide losses from conventional cropping: we have a body of knowledge, a context for CT-conventional tillage comparisons. Groundwater, however, is another story, another dimension of the problem about which we know much less.

THREE SUB-ISSUES

The above question can be broken down into three questions, listed below in order of increasing difficulty. These provide a research agenda.

Question 1: Does CT adoption increase pesticide use?

The answer is "yes" for herbicides, and "no generalization is possible" for other pesticides. Half of the conventional tillage used is for weed control (McWhorter and Shaw, 1982). Indeed, the development of chemical weed control and CT have been symbiotic all along (Wiese, 1985). Estimates for increases in herbicide use with CT depend on the practice, crop, and weed pressure, and range from 15 to 60% (Baker and Laflen, 1983; Baker et al., 1982; Crosson, 1981; Hanthorn and Duffy, 1984; Hinkle, 1983). All else being equal, we should expect proportionally higher herbicide losses in surface runoff from CT fields, as compared to conventionally-tilled crops (Wauchope 1978; Wauchope et al., 1985). All else is not usually equal, however. Fawcett (Chapter 2) maintains, for example, that in many areas the simultaneous development of CT and new herbicide uses have complemented each other, and the change has led to changes in kinds, but not total amounts, of pesticides applied.

Whether usage increases with CT for pesticides other than herbicides is less clear, and seems to be less of an issue, as indicated by Wagenet (Chapter 11). Why is this so?

Question 2: How does CT adoption affect the processes controlling pesticide runoff losses?

The reason this question is stated this way is that if we know the answer to question 1, then the answer to question 2 is the rest of what we need to know, to define the relative effects of CT *vs.* conventional practices on pesticide runoff.

To discuss what is known about question 2 we have to subdivide CT and distinguish between RT, practices, which can include significant soil disturbance, and NT, an extreme in which the only soil disturbance is cutting and closing a planting slot. Erosion control with NT is nearly total. NT results in a layered structure consisting of a live or dead mulch, below which is topsoil high in organic matter. The infiltration rate depends on the natural subsoil hydraulic conductivity (Edwards and Amerman, 1984). NT may actually *increase* runoff volume compared to conventional tillage (Lindstrom et al., 1984), but on average NT and conventional tillage have similar runoff volumes (Baker and Laflen, 1983). Sediment concentrations will be very low for NT.

The suggested "minimum" 30% crop residue definition for conservation tillage (Allmaras and Dowdy, 1985) means that nearly any operation except the moldboard plow is allowed. A 30% crop residue cover gives about 50% erosion control, and we can try to relate pesticide effects to these two figures. As for runoff/infiltration effects we can only suggest that a budget approach considering the interrelationships between storage, evapotransporation, runoff, and percolation below the root zone will have to examined in each case--not many combinations have been examined in the field (Gold and Loudon, 1982; Wauchope, 1986). Baker and Laflen (1983) suggest an average reduction in runoff volume of about 25% for RT.

Given these generalizations about the effects of RT and NT, we can make the following generalizations about their effects on pesticides in surface runoff under worse-case weather conditions.

NT will reduce losses of sediment-carried pesticides. This means any nonionic pesticide with a solubility less than about one mg/liter (Chapter 11; Wauchope, 1978), or pesticides with strong soil-binding mechanisms such as glyphosate or paraquat. Even if a NT practice requires a substantial increase in usage of such pesticides, the near-complete erosion control of NT should more than compensate. NT will probably increase losses of water-carried pesticides in surface runoff--possibly drastically.

NT may increase pesticide losses in the water phase of runoff--possibly drastically. If NT has little effect on runoff volume and more pesticide is used, we may expect higher losses as compared to conventional tillage. Much higher losses might occur if heavy herbicide use on mulch or cover crop is combined with increased runoff volume. RT will show somewhat lower losses of sediment-carried pesticides.

RT will have similar effects as NT, but to a lesser degree. Some increased pesticide use, combined with about a 50% erosion decrease, should about cancel each other, probably in favor of decreased losses. RT will probably give the same or slightly higher losses of water-carried pesticides. If use increases average 30-50% and runoff volume decreases average 25%, the average result is a small increase.

We are very simplistically saying that losses are proportional to use and that sediment and water, and their associated pesticides, can be treated independently. There are excellent field studies that broadly confirm this (Baker and Johnson, 1979; McDowell et al., 1981; Schwab et al., 1973; Smith et al., 1978; Triplett et al., 1978). It is important to remember also that many tillage operations and pesticide applications occur at the same time, at or near planting. Tillage effects on runoff volume and sediment load thus will occur during that period when applications are fresh and have the highest potential for leaching or runoff (Wauchope, 1978).

Fine-tuning these generalizations will require, as Wagenet has said (Chapter 11), the ability to account for several complicating factors. (1) The increased crop residue mulch or cover crop foliage under CT can intercept sprays (one reason for higher rates) and become an important source of pesticides in runoff (Deuel et al., 1977; Hall et al., 1984; Martin et al., 1978; Trichell et al., 1968). Conversely, mulch and grass cover can remove pesticides from runoff (Hall et al., 1983; Rhode et al., 1980; Trichell et al., 1968). Baker et al., (1982), however, got similar runoff of three herbicides whether they were placed above or below mulch. (2) A recent developement in herbicides is a series of diphenylether and sulfonylurea postemergence herbicides which are active at very low rates. Their use may lead to effective weed control with less chemical usage, even in CT systems. (3) New microencapsulated herbicide formulations designed to penetrate mulch in CT have unknown runoff characteristics. (4) Finally, the actual physical process of pesticide entrainment in runoff needs to be understood. For example, the first 0.5 cm of runoff removed as much as the remaining 3 cm in one study (Martin et al., 1978); in a tilted-bed simulation, 20% of the losses of water-carried pesticides were removed in the first 5% of the runoff water (Wauchope, 1987b). In these cases (see also White et al., 1967) pesticide concentrations are completely uncorrelated with sediment concentrations, in contrast to the high correlation that occurs with sediment-carried chemicals (McDowell et al., 1981; Wauchope, 1987a; White et al., 1967).

<u>Question 3: Does the increased pesticide usage in CT combine with runoff decreases to shunt pesticides to groundwater?</u>

Tillage practices can increase the fraction of rainfall that infiltrates instead of running off (Edwards and Amerman, 1984; Gold and Loudon, 1982; Triplett et al., 1968), and the result may be more downward mobility of pesticides. I feel that

Wagenet (Chapter 11) understates the amount of research that has been done on pesticide leaching. Burnside et al. (1963) showed 25 years ago that monuron, simazine and atrazine could be leached to subsoil areas where cooler soil temperatures and decreased microbial activity greatly increased their persistence. Other studies include picloram (Glass and Edwards, 1974) and triazines (Hall and Hartwig, 1978; Muir and Baker, 1976). What happened in many other cases, however (e.g., Sheets et al., 1972; White et al., 1976), was that compounds were found to be quite immobile, and this provided a false sense of security, later to be shattered by the aldicarb and EDB cases. Degradation products have proved more mobile than the parent compound in the cases of aldicarb, fenamiphos (Lee et al., 1986), and atrazine and cyanazine (Muir and Baker, 1976).

The aldicarb/EDB cases occurred in soils with very high hydraulic conductivities and where runoff seldom occured. Where CT drastically effects downward water movement, its use may concern us.

RESEARCH NEEDS

There are three approaches to sorting out the complexities of runoff/percolation and the associated pesticide loads as affected by CT. Each has particular strengths to offer to some of the research areas implied above.

Field Experiments

These are irreplaceable simply because the soil/weather/crop/pesticide system is so complex that we have to examine our assumptions and the variability of the system with real measurements. We need data in situations where the hydrology (both surface and subsurface) is well characterized, as in the Plains, GA project (Wauchope, 1986), and we need spatial variability studies such as done by Jury et al. (1986). We simply don't know enough about the subsoil region. We also need more multiresidue studies where different chemicals can be compared such as in the study by Wu et al. (1983) comparing atrazine and alachlor.

Physical Simulation

Using simulated rainfall and tilted-bed lysimeters (Wauchope, 1987a,b), one can answer many questions about the relative effects of formulations, soil properties, mulches, cover crops, and new chemicals, and the physical details of the runoff/leaching process. The best example of this is Sharpley et al., (1981) and their examination of the "effective depth of interaction" of rainfall on soil.

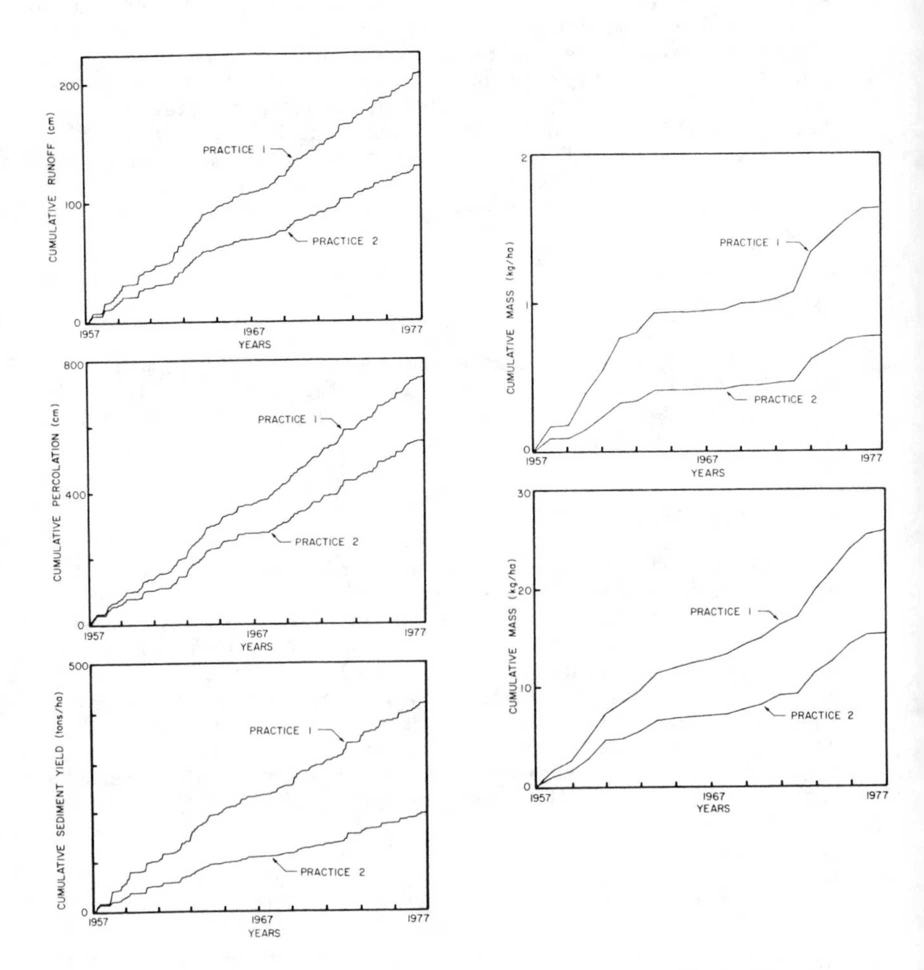

Figure 1. Calculated 20-year cumulative losses of runoff water, percolation water, sediment, and pesticides in runoff (upper right) and in percolation water (lower right), under conventional tillage (practice 1), and conservation tillage (practice 2). Adapted from Knisel (1985).

Mathematical Modeling

This technology has only begun to hit stride for agriculture and has potential limited only by our vision. We do modeling in part because we so often have to make management decisions without benefit of all the data, and in part because we can model complexity easier than we can characterize it by experiment. Weather profoundly effects all our predictions yet is itself stochastic, and modeling's greatest strength is to examine possibilities. For example, Knisel (1985) presents an example in which the chemical runoff model CREAMS is combined with 20 years of real weather data to calculate and compare the cumulative long-term pesticide runoff of conventional tillage (moldboard plow giving a hardpan at 20 cm) and RT systems (e.g., chisel to 30 cm plus residue plus rye winter cover crop) in continuous corn. The results provide several insights which would not be obtained any other way (Fig. 1). The simulation shows that chiseling gives less runoff and less leaching because of greater storage and evapotranspiration from the greater root depth. Pesticide losses both in surface and subsurface water differ by a factor of two.

These results also show that cumulative difference in pesticide losses even of 100% may not be distinguished over shorter time periods, depending on the weather pattern (particularly proximity of runoff to application each year).

CREAMS and other nonpoint pollution models are time-step-based physical models. Two other types of models are less explored which offer complementary aid. "Expert systems" provide an easier translation of our expertise into prediction systems--these are just beginning to be explored in our field. "Combinatory data base" systems, in which very simple physical models are combined with geographical information, may be used for comparisons of many areas of a selected region using data from the areas. An example is the DRASTIC system for predicting groundwater vulnerability (Aller et al., 1985). These are certain to become important decision-aiding tools in the future.

REFERENCES

Agrichemical Age: Tillage for the Times. 1984. International Minerals and Chemical Corp., 421 E. Hawley St., Mundelein, IL 60060.

Aller, L., T. Bennett, J. H. Lehr, and R. J. Petty. 1985. DRASTIC: a standardized system for evaluating ground water pollution potential using hydrogeologic settings. U. S. Environ. Prot. Agency R. & D. Rept. No. EPA/600/2-85/018, May 1985.

Allmaras, R. R. and R. H. Dowdy. 1985. Conservation tillage systems and their adoption in the United States. Soil & Tillage Res. 5: 197-122.

Baker, J. L. and H. P. Johnson. 1979. The effect of tillage systems on pesticides in runoff from small watersheds. Trans. ASAE 22: 554-559.

Baker, J. L. and J. M. Laflen. 1983. Water quality consequences of conservation tillage. J. Soil Water Conservation 38: 186-193.

Baker, J. L., J. M. Laflen, and R. O. Hartwig. 1982. Effects of corn residues and herbicide placement on herbicide runoff losses. Trans. ASAE 25: 340-343.

Burnside, O. C., C. R. Fenster, and G. A. Wicks. 1963. Dissipation and leaching of monuron, simazine and atrazine in Nebraska soils. Weeds 11: 209-213.

Crosson, P. 1981. Conservation tillage and conventional tillage: a comparative assessment. Soil Conservation Society of America, Ankeny, IA.

Deuel, L. E., K. W. Brown, F. C. Turner, D. G. Westfall and J. D. Price. 1977. Persistence of propanil, DCA, and TCAB in soil and water under flooded rice culture. J. Environ. Qual. 6: 127-132.

Edwards, W. M. and C. R. Amerman. 1984. Subsoil characteristics influence hydrologic response to no-tillage. Trans. ASAE 27: 1055-1058.

Glass, B. L. and W. M. Edwards. 1974. Picloram in lysimeter runoff and percolation water. Bull Environ. Contam. Toxicol. 11: 109-112.

Gold, A. J. and T. L. Loudon. 1982. Nutrient, sediment and herbicide losses in tile drainage under conservation and conventional tillage. ASAE 1982 Meeting, Paper No. 82-2549.

Hall, J. K., and N. L. Hartwig. 1978. Atrazine mobility in two soils under conventional tillage. J. Environ. Qual. 7: 63-68.

Hall, J. K., N. L. Hartwig, and L. D. Hoffman. 1983. Application mode and alternate cropping effects on atrazine losses from a hillside. J. Environ. Qual. 12: 336-340.

Hall, J. K., N. L. Hartwig, and L. D. Hoffman. 1984. Cyanazine losses in runoff from no-tillage corn in 'living' and dead mulches vs. unmulched, conventional tillage. J. Environ. Qual. 13: 105-110.

Hanthorn, M. and M. Duffy. 1984. Tillage choice and pesticide use in corn and soybeans. Agrichemical Age, February, 1984, 45-48.

Hinkle, M. K. 1983. Problems with conservation tillage. J. Soil Water. Conserv. 38: 201-206.

Jury, W. A., G. Sposito, and R. E. White. 1986. A transfer function model of solute transport through soil. 1. Fundamental concepts. Water Resources Res. 22: 243-247.

Knisel, W. G. 1985. Use of computer models in managing nonpoint pollution from agriculture. Nonpoint Pollution Abatement Symposium, Milwaukee Wisconsin, April 23-25, 1985.

Lee, C.C., R. E. Green, and W. J. Apt. 1986. Transformation and adsorption of fenamiphos, F. sulfoxide and F. sulfone in Molokai soil and simulated movement with irrigation. J. Contaminant Hydrology 1: 211-225.

Lindstrom, M. J., W. B. Voorhees, and C. A. Onstad. 1984. Tillage system and residue cover effects on infiltration in northwestern corn belt soils. J. Soil and Water Conservation 39: 64-68.

Martin, C. D., J. L. Baker, D. C. Erbach, and H. P. Johnson. 1978. Washoff of herbicides applied to corn residue. Trans. ASAE 21: 1164-1168.

McDowell, L. L., G. H. Willis, C. E. Murphree, L. M. Southwick and S. Smith. 1981. Toxaphene and sediment yields in runoff from a Mississippi Delta watershed. J. Environ. Qual. 10: 120-125.

McWhorter, C. G. and W. C. Shaw. 1982. Research needs for integrated weed management. Weed Sci. 30 (Suppl.): 40-45.

Muir, D. C., and B E. Baker. 1976. Detection of triazine herbicides and their degradation products in tile-drain water from fields under intensive corn (Maize) production. J. Agric. Food Chem. 24: 122-125.

Rhode, W. A., L. E. Asmussen, E. W. Hauser, R. D. Wauchope, and H. D. Allison. 1980. Trifluralin movement in runoff from a small agricultural watershed. J. Environ. Qual. 9: 37-42.

Schwab, G. O., E. O. McLean, A. C. Waldron, R. K. White, and D. W. Michener. 1973. Quality of drainage water from a heavy-textured soil. Trans. ASAE 1973: 1104-1107.

Sharpley, A. N., L. R. Ahuja, M. Yamamoto, and R. G. Menzel. 1981. The kinetics of phosphorus desorption from soil. Soil Sci. Soc. Amer. J. 45: 493-496.

Sheets, T. J., J. R. Bradley, and M. D. Jackson. 1972. Contamination of surface and groundwater with pesticides applied to cotton. Univ. NC WRRI Rept. No. 60, April 1972, 63 pp.

Smith, C. N., R. A. Leonard, G. W. Langdale and G. W. Bailey. 1978. Transport of agricultural chemicals from small upland Piedmont Watersheds. U.S. Environmental Protection Agency Res. Rept. EPA-600/3-78-056, May 1978.

Trichell, D. W., H. L. Morton and M. G. Merkle. 1968. Loss of herbicides in runoff water. Weed Sci. 16: 447-449.

Triplett, G. B., B. J. Connor, and W. M. Edwards. 1978. Transport of atrazine and simazine in runoff from conventional and no-tillage corn. J. Environ. qual. 7: 77-84.

Triplett, G. B., D. M. Van doren, and B. L. Schmidt. 1968. Effect of corn stover mulch on no-tillage corn yield and water infiltration. Agron. J. 60: 236-239.

Wauchope, R. D. 1978. The pesticide content of surface water draining from agricultural fields: a review. J. Environ. Qual. 7: 459-472.

Wauchope, R. D. 1986. The Plains Project: An integrated study of the impacts of agriculture on surface and groundwater quality in the Georgia Coastal Plain. p. 25-37 IN A. R. Bonanno and T. J. Monaco (eds.), Pesticide Groundwater Contamination--Is It Real? Proc. NCWSS, Raleigh, NC.

Wauchope, R. D. 1987a. Tilted-bed simulation of erosion and chemical runoff from agricultural fields. 1. Description of apparatus and results with runoff of sediment and sediment-associated copper and zinc. J. Environ. Qual. (in press).

Wauchope, R. D. 1987b. Tilted-bed simulation of erosion and chemical runoff from agricultural fields. 2. The effects of formulation on atrazine runoff. J. Environ. Qual. (in press).

Wauchope, R. D., L. L. McDowell, and L. J. Hagen. 1985. Environmental effects of Limited Tillage. P. 266-281 In A. F. Wiese (Ed.), Weed Control in Limited-Tillage Systems, Weed Sci. Soc. Amer., Champaign, IL.

White, A. W., L. E. Asmussen, E. W. Hauser, and J. W. Turnbull. 1976. Loss of 2,4-D in runoff from plots receiving simulated rainfall and from a small agricultural watershed. J. Environ. Qual. 5: 487-490.

White, A. W., A. P. Barnett, B. G. Wright, and J. H. Holladay. 1967. Atrazine losses from fallow lands caused by runoff and erosion. Environ. Sci. Technol. 1: 740-744.

Wiese, A. F. (Ed.) 1985. Weed Control in Limited-Tillage Systems, Weed Sci. Soc. Amer., Champaign, IL.

Wu, T. L., D. L. Correll, and H. E. H. Remenapp. 1983. Herbicide runoff from experimental watersheds. J. Environ. Qual. 12: 330-336.

CHAPTER 13

EFFECT OF CONSERVATION TILLAGE ON FATE AND TRANSPORT OF NITROGEN

J. W. Gilliam and G. D. Hoyt,
North Carolina State University, Raleigh, North Carolina

INTRODUCTION

A review of the literature with regard to nitrogen (N) dynamics in the conservation tillage system reveals that some processes have been studied fairly extensively and others very little. Our approach in this paper will be to utilize what is known about the fate of N in some processes to estimate the quantities of N which may be involved in others studied less. Crop utilization efficiencies, denitrification, immobilization and losses by runoff are examples of processes on which there is sufficient information to reach reliable conclusions about the effects of conservation tillage on nitrogen. The quantity of N that is leached or volatilized in a conservation tillage system is less well known. With a few notable exceptions, most researchers have obtained these latter values by difference and we will also.

VOLATILIZATION

There is no question that volatilization of N in no-till (NT) is a problem, particularly when solid urea or urea-ammonium nitrate are used as fertilizer sources. There have been little or no direct measurements of quantities of N lost but a decreased efficiency of surface applied urea as compared to other sources of N (Mengel et al., 1982; Bandel et al., 1980; Touchton and Hargrove 1982; Carter and Rennie, 1984) is believed to be largely a result of gaseous loss. Wells (1984) concluded,

Effects of Conservation Tillage on Groundwater Quality: Nitrates and Pesticides, Terry J. Logan et al., eds.

as we do, that Bandel et al., (1980) put the loss of N from surface applied urea in NT fields in proper perspective when they concluded that a quantitative estimate of the loss of ammonia can seldom be predicted with reasonable certainty. There is no evidence that indicates that volatilization of N from other N sources is a problem in NT.

Because of the increasing importance of urea as a N fertilizer source, volatilization losses of N applied as urea is an area of concern. Touchton and Hargrove (1982) observed lower losses when N solutions were banded on the surface as compared to broadcast but they concluded that urea was not efficient for surface applications to NT systems. Several of the above mentioned studies have indicated no difference in efficiency of urea as compared to other sources when the fertilizer was incorporated. Baker and his associates at Iowa State have worked with various designs for a point-injector applicator that incorporates liquid fertilizer without residue incorporation (Baker and Laflen, 1983). Incorporation of N fertilizers is an area requiring more work if volatilization losses are going to be held to acceptable levels. Surface applications of urea-based fertilizers will result in a quantitatively unpredictable loss of N as ammonia.

DENITRIFICATION

Denitrification studies on NT cropping systems have occurred under various crop rotations and climatic regimes in the United States, Great Britain and Canada (Table 1). Groffman (1985) measured denitrification rates on a monthly basis from a Typic Rhodudult soil from Georgia and concluded that when soil moisture was low, NT soils would lose more nitrogen to denitrification than conventionally tilled soils because of the higher moisture levels in the NT soils, but that both systems would produce low rates (<10 kg N/ha/yr). Linn and Doran (1984b) also concluded that greater water filled pores created by the NT system favored up to 9.4 times higher N losses through denitrification than conventional tillage. They also concluded that nitrous oxide production would be higher in NT under both wet and dry conditions. Dowdell et al., (1983) concluded that wet soils in Great Britain would lose more N to denitrification in direct-drilled (NT) management than in plowed systems.

Rice and Smith (1982) investigated denitrification losses from a well drained soil series (Typic Paleudalfs) in Kentucky and concluded, like other researchers, that higher denitrification activity was found in soils taken from NT culture than from conventionally tilled areas. They also stated that the higher moisture in the NT and not an increase in energy source would be the primary reason for an increase in denitrification. In later experiments in Kentucky (Kitur et al., 1984), little or no denitrification was observed in a three year field experiment. This led Wells (1984) to conclude that

Table 1. The Effect of Tillage, Cropping System, Soil Moisture, and Location on Denitrification

Tillage	Cropping System	Soil Moisture	Location	Soil	Potential Denitrification Loss	Leaching & Denitrification	Reference
NT>CT	Rye/sorghum	Low	Georgia	Rhodudolt	< 10 kg N/ha/yr		Groffman, 1985
NT=CT	Rye/sorghum	High	Georgia	Rhodudult	> 25 kg N/ha/yr		Groffman, 1985
NT>CT	Wheat	High	Great Britian				Dowdell et al., 1983
NT>CT	–	High	Kentucky	Hapludalf			Rice et al., 1982
NT	Rye/sorghum		Georgia	Rhodudult		20 kg N/ha/yr	House et al., 1984
CT	Rye/sorghum		Georgia	Rhodudult		24 "	"
NT	Rye/soybeans		Georgia	Rhodudult		26 "	"
CT	Rye/soybeans		Georgia	Rhodudult		32 "	"
NT	Field Crops		Saskatoon, Canada	Haploboroll	12–16 "		Aulakh et al., 1984
CT	Field Crops		Saskatoon, Canada	Haploboroll	3–7 "		"
NT	Fallow		Saskatoon, Canada	Haploboroll	34 "		"
CT	Fallow		Saskatoon, Canada	Haploboroll	12–14 "		"

CT is Conventional Tillage
NT is No Tillage

denitrification was of minor importance in NT soils in Kentucky; also when differences were observed, that denitrification in NT would be greater.

Aulakh et al., (1984) measured denitrification rates in conventionally tilled and NT plots cropped to wheat or fallow on Typic Haploboroll soils. Cropped fields showed low levels of denitrification with zero-tilled twofold higher than conventional systems. Fields that were previously in conventionally tilled or NT wheat with a fallow rotation were higher in denitrification than continuous wheat. Gaseous N losses were related to the higher soil density and greater moisture in the NT plots.

House et al., (1984) calculated leaching and denitrification losses from a CT and NT rye/sorghum and rye/soybean rotation and concluded that losses from these two processes would range from 20-32 kg N/ha/yr with the smaller losses occurring in the NT system. These smaller losses were a result of greater immobilization in the NT system.

Microbial populations measured from six midwestern U.S. locations confirmed the potential increase in denitrifiers from NT surface soils over conventionally tilled soils. Linn and Doran (1984a) reported 1.27 to 1.31 times greater anaerobic microorganisms in the 0 to 7.5 cm soil depth from NT soils. They attributed the presence of less aerobic conditions in the NT to the greater denitrification potential. Doran (1980) surveyed various locations throughout the U.S. (Oregon, Nebraska, Minnesota, Kentucky, and West Virginia) for differences in denitrifier populations and concluded that NT soils in the 0 to 7.5 cm depths contained up to 7 times the microbial populations as CT soils. He also stated that facultative anaerobes and denitrifiers from the aerobic population were twice as high in the NT soils collected at depths of 7.5 to 15 cm. Broder et al., (1984) also support these findings by stating that both spring soil water content and cooler soil temperatures in an Aridic Arguistoll soil in Nebraska promoted a higher population of denitrifiers.

Our conclusion from the available literature is that there is greater potential for denitrification in the NT system. Under very wet conditions where denitrification is high, there is probably little effect of tillage system. However, under conditions which are marginal for denitrification to occur, NT areas are likely to lose more nitrogen to denitrification. It would be very difficult to defend a quantitative estimate of what this average increase in denitrification might be but our guess is 2-8 kg N/ha/yr.

NITRIFICATION

There has been little work done on nitrification rates under conservation tillage systems. Rice and Smith (1982) concluded in one study that there was no difference between tillage systems in nitrification rates as long as moisture remained the same. They did measure increased rates of nitrification in NT Kentucky field soils because of better moisture conditions. They concluded that this frequently would occur because dry conditions would not limit nitrification as often in NT soils. In another study, they measured higher ratios of nitrate to ammonium in CT soils than in NT soils (Rice and Smith, 1983). They found nitrification rates similar for the CT and NT soils, but the higher moisture levels in the NT promoted greater production of nitrate in fertilized soils. In contrast, NT soils that received no fertilizer additions had a lower mineralization of ammonium and consequently less nitrification. Broder et al., (1984) measured potential mineralizable N (PMN) in fertilized NT and CT soils in Nebraska and found that the NT soils contained higher PMN. They suggest that the NT soils may have immobilized more soluble N or mineralized less N during the previous cropping sequence and thus produced lower nitrate concentrations in the soil. A parallel reduction in nitrifier population was found.

Doran (1980) also suggested that soil water was the primary environmental factor affecting microbial populations between the two tillage systems. He collected samples from seven locations throughout the U.S. under various cultural and crop managements and measured higher numbers of ammonium oxidizers from the 0 to 7.5 cm layer of NT soils. He related NT ecosystems to those that are undisturbed such as grasslands. Groffman (1985) concluded that there was more nitrification in the 0 to 5 cm depth in NT soils, less nitrification in the 13 to 21 cm depth with the net effect of no difference on an annual basis. Dowdell and Cannell (1975) measured mineral N from soil cores taken from a field location in England that had been plowed or direct drilled with spring cereals for several years. Concentrations of nitrate during the winter at the 30 cm depth were 2 to 5 times greater after plowing than in direct drilled soils. There were no differences between the two tillage systems in mineral N when measurements were made in the spring.

Our conclusion is there may be slightly more N available for leaching in the NT system because of increased nitrification of added N during the early spring but the difference between tillage systems with regard to nitrification is likely to be small.

NITROGEN IMMOBILIZATION AND MINERALIZATION

Stanford et al., (1973) reviewed initial studies of nitrogen movement in conservation tillage and found organic nitrogen levels higher in mulch-till or NT treatments than in conventional tillage. They indicated from the work of Beale et al., (1955) that both organic matter and nitrogen accumulated faster in mulch-till than in CT. Doran and Power (1983) concluded that higher moisture levels created by a NT surface mulch produced a less oxidative biochemical environment and produced a greater concentration of plant roots in the soil surface layer. These enchanced microbial populations and provided a greater potential for soil immobilization of N in the humid regions of the U.S.

Free (1970) established a seven year NT corn experiment on a Honeoye silt loam soil in New York and found that NT increased immobilization up to 122 kg N/ha/yr as compared to the CT treatment. After five years of comparison of NT and CT corn on a Cecil clay loam soil (Typic Hapludult) in Virginia, Moschler et al., (1972) measured higher soil organic N (3.6 kg N/cm soil/ha/yr) in NT plots sampled in 15 cm. A similar gain of organic N was also found in a Davidson soil (Rhodic Paleudult) cropped for six years. On a Lodi silt loam soil (Typic Hapludult) slightly higher organic N was found in the CT treatments than in NT. They attributed weather conditions for the difference in N immobilization between the Cecil and the Lodi soil.

Although soil nitrogen fluctuated substantially in both NT and CT rye/sorghum yield plots (0-10 cm) in Georgia, organic N remained greater in the NT soil than in the CT soil after a 2 year cropping sequence (Stinner et al., 1983). Both NT and CT soil organic N followed similar flow patterns as an adjacent old-field site throughout the sampling period but remained lower than the old-field location. After four years, lower N flux rates indicated an increase in immobilization and higher N retention under the NT cropping system (House et al., 1984). Both organic C and organic N increased significantly with increasing N rates under NT than CT corn in Kentucky (Blevins et al., 1977). Moderate rates of N left the soil chemical characteristics similar to those originally found in bluegrass sod five years earlier, suggesting that intensively cropped NT soils can maintain or preserve conditions similar to native conditions. Bandel et al., (1975) established tillage experiments on three soils similar in soil taxonomy and found residual mineral N in the root zone similar in both the tilled and untilled treatments. Organic N was greater in two soils in the NT treatment and lower in one soil.

In an erosion-prone tropical Alfisol in western Nigeria, Lal (1976) measured slower decline of soil organic C and more organic N (.10 and .07 % N, for NT and CT soils, respectively) in NT fields cropped five years to various legume-grass combinations (maize, cowpeas, soybeans, and pigeonpeas) than in CT systems. In a Typic Hapludalf soil in Florida, Ferrer et al., (1984) found higher organic N levels (.065 and .047 % N, for NT and CT soils, respectively) in the surface soils (0-5 cm) of NT corn fields cropped continuously to corn than those under conventional culture, with no apparent difference in the soil C/N ratio. Under similar climatic conditions, Ortiz and Gallaher (1984) also found higher levels of organic N in a NT Grossarenic Paleudult soil than in a CT soil in Florida, but state that the C/N ratio of the soils were lower in the NT field (13.4 and 17.0 for NT and CT soils, respectively), creating a higher potential for mineralization from the NT soils.

In two year old wheat-fallow/wheat-wheat rotations in Saskatchewan, Canada, Auakh et al., (1984) observed higher organic N (18.5 Kg N/Cm soil/ha/yr) from the 0-29 cm soil surface of NT over CT field plots. Fall applied nitrogen on an Elstow clay loam soil in Saskatchewan was sampled the following May for % N recovery (Aulakh and Rennie, 1984). No-tilled stubble soils immobilized 15.2% of the applied N into the organic N pool (0-45 cm) while 51.4% was recovered as mineral N. Losses of N not recovered were thought to be removed by dentrification. Although higher recovery of N was found in the CT stubble soils (74.8% vs 66.6% for CT and NT, respectively), the NT stubble soils maintained similar levels of mineral N (51.4% and 54% for CT and NT, respectively).

El-Haris et al., (1983) studied the effect of long-term crop rotations (seven years) and tillage on various wheat, pea, alfalfa, and fallow management systems in Washington State. In samples taken in the fall from a Palouse silt loam soil, uniform nitrogen mineralization potential values were measured throughout the entire 0-15 cm depth of a plowed soil, whereas, in the NT treatments the 0-5 cm depth was higher in mineralization potential than the 5-15 cm depth. This effect gave a net value that was similar to the CT treatments for this 0-15 cm depth. Spring sampling showed a higher level of nitrogen mineralization potential in the NT than in the CT treatments for this 0-15 cm depth. Conventional tillage did result in higher levels of mineral N, organic C, and a lower C/N ratio than in the NT treatments.

In long-term (25 years) stubble mulch soils established in the northern Great Plains in 1954, Bauer and Black (1981) reported a nitrogen loss of 23% from stubble mulched soils and 33% from conventional-tilled soils. They suggest higher rates of erosion and greater oxidation potential from the CT field locations as the main reasons for the difference.

In a fall-killed fallow soil experiment, Arnott and Clement (1966) measured ammonium and nitrate concentrations from the surface 0-15 cm of herbicide-treated and plowed field plots. They found a progressive increase in mineralizable N for both treatments, but with the plowed soil having slightly higher total concentrations of mineral N. They state that these final higher concentrations of mineral N resulted from an earlier release of mineral N from the plowed soils before the herbicide treated soils started to release mineral N. Spring-killed fallow soil and plowed soils followed a similar pattern as the fall-killed and plowed experiment, except that the levels of mineralization were higher in the spring. Again, the plowed soils started to release mineral N earlier after treatment (plowing or herbicide kill) than did the herbicide killed treatment.

Numerous soil N studies with direct drilling can be found in Fleige and Baeumer (1974) (Table 2) which describes relative nitrogen changes due to tillage, but in general, does not provide actual N (kg N/ha) for use in input-output modeling. In these studies, organic N either mineralized or immobilized at different rates and information of the effect of tillage is limited for comparison.

TABLE 2. MEAN ANNUAL INCREASE IN SOIL ORGANIC NITROGEN OF ARABLE SOILS (0-15 cm DEPTH) AS CAUSED BY ZERO- TILLAGE.

Soil, location	Continuance (years)	% total N in dry soil	Reference
Sand			
Muncheberg, Germany Democratic Republic	4	0.0015	Buhtz et al. (1970)
Silt loam			
Gottingen 3, German Federal Republic	6	0.0009	
Gottingen 2, German Federal Republic	5	0.0015	
Gottingen 1, German Federal Republic	5	0.0021	
Koetschau, German Democratic Republic	4	0.0030	Buhtz et al. (1970)
Seehausen, German Democratic Republic	4	0.0050	Buhtz et al. (1970)
Stuttgart 1, German Federal Republic	3	0.0081	Kahnt (1971)
Loam to clay loam			
Trossin, Germany Democratic Republic	4	0.0005	Buhtz et al. (1970)
Stuttgart 2, German Federal Republic	3	0.0081	Kahnt (1971)
Stuttgart 3, German Federal Republic	3	0.0117	Kahnt (1971)

From Fleige and Baeumer, 1974.

There are, however, a number of studies with sufficient data provided to calculate N immobilized or mineralized. Information from some experiments is available to quantify organic nitrogen transformations from when the study was initiated to a time period after various cropping seasons when additional samples were taken (Table 3). Some experiments have introduced conservation-tillage on soils that were previously under cultivation (Fleige and Baeumer, 1974; Doran and Power, 1983), while others have initiated a long-term conservation tillage experiment from old-field (Stinner et al., 1983) or

Table 3. The Effect of Changing Tillage Practice on Gain or Loss of Soil Organic Nitrogen

Tillage	Years	Location	Soil	Depth cm	Nitrogen, gain(+) or loss(-) kg N/cm soilha/yr	kg N/ha/yr	Reference
CT to NT	1	Georgia	Tifton Plinthic Paleudult	0-10	+2.4	+24	Hoyt, unpublished data
CT to NT	5	Gottingen, G.F.R	Grey-Brown Podsolic	0-30	+2.4	+72	Fleige & Baeumer, 1974
CT to NT	5	Gottingen, G.F.R.	Grey-Brown Podsolic	0-30	+2.2	+66	Fleige & Baeumer, 1974
CT to NT	6	Gottingen, G.F.R.	Grey-Brown Podsolic Aridic Argiustoll	0-30	+1.8	+54	Fleige & Baeumer, 1974
CT to CT	10	Nebraska	"	0-15	-0.4	-6	Doran & Power, 1983
Oldfield to NT	2	Georgia	Hiwassee loam Typic Rhodudult	0-10	-5.3	-53	Stinner et al., 1983
Oldfield to CT	2	Georgia	"	0-10	-6.0	-60	Stinner et al., 1983
Grassland to NT	5	Gottingen, G.F.R.	Grey-Brown Podsolic	0-30	-5.4	-162	Fleige & Baeumer, 1974
Grassland to NT	9	Nebraska	Duroc loam Pachic Haplustoll	0-15	-0.7	-11	Doran & Power, 1983
Grassland to CT	9	Nebraska	"	0-15	-3.0	-45	Doran & Power, 1983

grassland locations (Fleige and Baeumer, 1974; Doran and Power, 1983; Blevins et al., 1977; Blevins et al., 1983). Results from these various studies located throughout the U.S. and Europe show strikingly similar trends, in that going from previous conventional culture to NT will result in an increase in organic N (immobilization) from 1.4 to 2.4 kg N/cm soil/ha/yr for the range of 0-30 cm surface soil. Under conditions where these studies have been placed in soils that were previously undisturbed (grassland or old-field), mineralization of the organic N has occurred for both conservation tillage and where companion conventional culture has been established. Mineralization rates for soils under conservation tillage have gone from 0.7 kg N/cm soil/ha/yr to as high as 7.2 kg N/cm soil/ha/yr. Mineralization rates for soils going from undisturbed to conventional culture range from 3 to 22.4 kg N/cm soil/ha/yr. In all cases, conventional culture accelerates soil organic N mineralization faster than does conservation tillage.

Experiments listed in Table 4 have compared the relative organic N transformed by mineralization (loss-) or immobilization (gain+). The net N gain or loss reflects the difference in tillage systems in ability to conserve nitrogen. These values represent changes that have occurred between the two tillage practices, regardless of previous cultural conditions before the study was initiated. For example, a two year experiment by Stinner et al., (1983) shows a net positive organic N gain of 4.1 kg N/cm soil/ha/yr by the use of conservation tillage over CT. The data in both tables reflect the slower rate of mineralization from the NT soil. An inverse of this N transformation occurred (Doran and Power, 1983) when the study was initiated from conventional culture (CT to CT) (Table 3). When this management practice is established, immobilization of organic N will occur faster in the NT soils than in continuous CT culture.

All of the data on mineralization and immobilization suggest that both CT and NT have an steady state level of soil N. Once this level is reached, there will be fluctuations in N present in the soil but no net changes until the management system is changed. Thus soil organic matter serving as a sink for applied N when management changes from CT to NT is temporary. This N reserve may serve as a buffer against N deficiencies but probably will have little long term effect on fate of applied N.

EFFICIENCY OF FERTILIZER N

Very little will be covered in this paper on this topic because other chapers in this book discuss it in more detail. Also, the topic was covered very well by Wells (1984) in a recent review. Thus, we will only give some general information which was utilized in estimating the availability of N for leaching in conservation tillage systems. At low rates of fertilizer N additions, conservation tillage usually initially

TABLE 4. THE EFFECT OF TILLAGE ON GAIN OR LOSS OF SOIL ORGANIC NITROGEN

TILLAGE	YEARS	LOCATION	SOIL SERIES	DEPTH cm	NITROGEN, gain(+) or loss(-) kg N/cm soil/ha/yr	kg N/ha/yr	REFERENCE
NT compared to CT	1	Maryland	Beltsville silt loam Typic Fragiudult	0-15	+1.0	+15	Bandel et al., 1975
NT compared to CT	1	Maryland	Bertie silt loam Aquic Hapludult	0-15	-2.8	-42	Bandel et al., 1975
NT compared to CT	1	Maryland	Mattapex silt loam Aquic Hapludult	0-15	+7.3	+110	Bandel et al., 1975
NT compared to CT	1	Georgia	Tifton Plinthic Paleudult	0-10	+2.1	+21	Hoyt, unpublished data
NT compared to CT	2	Georgia	Hiwassee loam Typic Rhodudult	0-10	+4.1	+41	Stinner et al., 1983
NT compared to CT	5	Georgia	Hiwassee loam Typic Rhodudult	0-21	+5.3	+111	Groffman, 1984
NT compared to CT	5	Virginia	Cecil clay loam Typic Hapludult	0-15	+3.6	+54	Moschler et al., 1972
NT compared to CT	5	Kentucky	Maury silt loam Typic Paleudalf	0-5	+15	+75	Blevins et al., 1977
Grassland compared to NT	5	Kentucky	Maury silt loam Typic Paleudalf	0-5	-7.2	-36	Blevins et al., 1977
Grassland compared to CT	5	Kentucky	Maury silt loam Typic Paleudalf	0-5	-22	-112	Blevins et al., 1977
ZT compared to CT	7	New York	Honeoye silt loam Glossoboric Hapludalf	0-7.6	+16	+122	Free, 1970
NT compared to CT	10	Kentucky	Maury silt loam Typic Paleudalf	0-5	+12	+60	Blevins et al., 1983
Grassland compared to CT	10	Kentucky	Maury silt loam Typic Paleudalf	0-5	-9.0	-45	Blevins et al., 1983
Grassland compared to NT	10	Kentucky	Maury silt loam Typic Paleudalf	0-5	-6.2	-31	Blevins et al., 1983
ZT compared to CT	25	North Dakota	Vebar Typic Haploboroll	0-45.7	+1.9	+87	Bauer & Black, 1981
ZT compared to CT	25	North Dakota	Grail Pachic Argiboroll	0-45.7	+1.0	+46	Bauer & Black, 1981

results in lower efficiency as compared to CT. This lower efficiency is probably a result of increased immobilization in conservation tillage at the lower N rates. Several researchers have measured either an increased N fertilizer efficiency for conservation tillage or no difference between tillage systems at the higher N rates. Increased efficiency is probably a result of better moisture conditions in conservation tillage. All of these factors tend to encourage higher rates of N applications in conservation tillage management systems.

Even when there is relatively high fertilizer N efficiency at high rates, there is still a large percentage of the fertilizer N which is not harvested with the grain. It is this N which is of environmental concern, particularily if it is not immobilized or denitrified.

LOSSES BY SURFACE RUNOFF

There have been several studies conducted on the effects of conservation tillage on losses of nitrogen in surface runoff water. Baker (1985) recently made an excellent comprehensive review of the available literature on this topic and concluded that there was qualitative agreement of the data obtained. There is not question that conservation tillage in general reduces runoff and losses of nitrogen via this route. The amount of the reduction in volume of runoff water has been variable both between locations and between years at the same location as seen in Table 5 adapted from Baker (1985). The values given in Table 5 are expressed as a percent of the values obtained for CT and range from 9 to 109%. Wendt and Burwell (1985) and Baker and Laflen (1983) both concluded that the average reduction was probably 20-25% and we agree that this estimate is as good as any if an average is desired.

The reduction in the amount of N lost in surface water as a result of utilization of conservation tillage has not been as great as the reduction in the amount of sediment. There is generally a higher concentration of both dissolved N in the surface drainage water and total N in the sediments from conservation tillage as compared to CT (Laflen and Tabatabai, 1984; Baldwin et al., 1985; Angle et al., 1984; Alberts et al., 1981; Baker and Laflen, 1982; McDowell and McGregor, 1980). The data in Table 5 adapted from Baker (1985) illustrate these points very well.

The higher average concentrations of dissolved N in the runoff result from several factors. The most important is that fertilizer N is usually surface applied without being incorporated (Wells, 1984; Timmons et al., 1970; McDowell and McGregor, 1980). Several researchers have observed little or no difference in runoff concentrations between tillage systems when nitrogen fertilizer was incorporated (Baker and Laflen, 1982; Timmons et al., 1970). Another factor contributing to the higher concentration is that soil N tends to be higher at the

Table 5. Runoff, Erosion and Nitrogen Losses for Conservation Tillage Expressed as a Percentage of Those for Conventional Tillage (Baker, 1985).

Practice	Study	Soil	Slope	Rnof	Eros	Nitrogen Slon	Nitrogen Sed't	Nitrogen Total	Ref.	Crop
			%	%	%	%	%	%		
Till-plant	N,W[1]	sil	10-15	65	38	68	54	55	2	cont.corn
No-till (ridge)				58	11	44	19	20	2	cont.corn
No-till	N,W	sil	9	9	1				3	cont.corn
No-till	N,P	sil	5	51	1	70	6	10	4	bean-bean
No-till				38	1	190	6	21	4	bean-wheat
No-till				106	12	180	40	52	4	bean-corn
Till-plant	N,P	sicl	6	71	58				5	corn
Till-plant	S,P	sil	8-12	86	33	2100	41	92	6	cont.corn (fert. trt.)
Till-plant	S,P	sil	5-12	83	77	200	68	70	7,8	cont.corn
no-till				90	17	120	21	22	9	corn
No-till				109	36				10	row crop

1 N indicates natural precipitation; S, simulated rainfall; W, watershed; and P, plot.
2 Johnson *et al.*, 1979. 3 Harrold *et al.*, 1970.
4 McDowell and McGregor, 1980. 5 Onstad, 1972. 6 Romkens *et at.*, 1973.
7 Barisas *et al.*, 1978. 8 Laflen, *et al.*, 1978.
9 Siemens and Oschwald, 1978. 10 Laflen and Colvin, 1981.

soil surface in the conservation tillage systems (House et al., 1984; Powlson and Jenkinson, 1981; Doran, 1980). Thus equilibrium concentrations of both organic and ammonium nitrogen would be higher. There is also some leaching of nitrogen from the plant residue remaining on the surface (McDowell and McGregor, 1980; Timmons et al., 1970) in the minimum till systems although the quantities of N losses via this pathway would not be expected to be large. Solution N losses measured from conservation tilled systems have ranged from 44 to 2100% (Table 5).

The above discussion indicates a very significant reduction in losses of organic nitrogen associated with sediment under conservation tillage with a concomitant increase in loss of solution nitrogen. Although solution N is more readily available to microorganisms contributing to eutrophication problems, the absolute increase in losses via surface runoff are usually not very large. Even with the large range of values measured, the expected average annual increase is probably 2 to 4 kg N/ha. Environmental problems resulting from losses of N from agricultural fields are almost always a result of subsurface movement of inorganic nitrogen (Gilliam, 1985; Gilliam et al., 1985) and conservation tillage is not likely to change that.

LOSSES OF NITROGEN BY LEACHING

In contrast to the data available on surface losses of nitrogen as influenced by tillage methods, there is very little information available on effect of leaching. Fortunately, the work which has been done is very definitive so some conclusions can be drawn. The work which has been most often quoted over the past few years was by Thomas and associates in Kentucky (Thomas et al., 1973; Tyler and Thomas, 1979; and Thomas et al., 1981). These workers noted that considerable more nitrogen leached below 90 cm in a killed sod NT system than in a CT treatment. The nitrogen which was leached in their system largely came from surface applied ammonium nitrate. The workers in Kentucky concluded that there was a potential for greater leaching in conservation tilled systems as compared to CT particularily during the early growing season. Their results are consistent with greater infiltration into a soil which was already at a higher moisture content and containing more continuous pores (Goss et al., 1978). Because essentially all work showed that conservation tillage resulted in greated infiltration, it has been accepted until recently that greater leaching would also occur. Recent work has shown that this assumption is not always correct.

Kanwar et al., (1985) working on a loam soil in Iowa observed much less leaching of N in NT plots than in moldboard plow plots. The difference between these observations in Iowa and the previously cited data from Kentucky can be explained. In both locations, much of the water movement was in large pores in the NT areas. This allowed the water containing the fertilizer N to move deeper in the soils in Kentucky so deeper N movement was observed. In Iowa, much of the nitrate was present in the soil before fertilization. Thus the water which moved in the large pores bypassed much of the N present in the profile so that less nitrate leaching occurred. This points out the importance of consideration of soil characteristics before any generalizations can be made.

More recent work in Kentucky (Kitur et al., 1984) indicated no difference between tillage systems in leaching of nitrogen. This work was conducted on soils very similiar to previous experiments where more leaching was observed. Thus many variables must be considered (soil type, fertilization method and rainfall) to predict relative movement of nitrogen through the profile, particularily for a particular year or growing season.

If we consider the total N balance for both the NT and CT systems, our previous conclusions would indicate that more leaching of N would occur. We have previously concluded that there is a greater chance of volatilization under conservation tillage but this problem is likely to be solved by better applicaton methods. The utilization efficiency by the crop is not greatly different when recommended rates of nitrogen are applied. There is frequently a higher concentration of nitrogen as soil organic N in the NT soils but this only indicates a higher equilibrium level of soil N and cannot be a continual sink for applied N. There is general agreement that less nitrogen is lost via surface runoff under CT. The difference between the amounts lost in surface runoff must either be denitrified or be lost to leaching waters if our other conclusions regarding N are correct. Thus either denitrification in NT must be increased by the amount of the decrease in runoff losses or leaching of N must increase. It is our opinion that leaching is likely to increase more than denitrification. There is essentially no long term direct information to support this conclusion, however. Gold and Loudon (1982) observed more nitrate-nitrogen leaving in tile drainage water from conservation tillage plots but their increase was only 1.6 kg/ha/yr. Considerably more information will have to be obtained before any definitive statements can be made about the long term effects of conservation tillage on leaching.

If leaching is increased as we believe, conservation tillage may have a net negative environmental effect with regard to nitrogen. It is clear that less organic N is lost with surface water but this decrease may be relatively unimportant because this is not the nitrogen which causes the greatest environmental problems. The N which is leached may enter the surface water at lower elevations and cause environmental problems (Chichester, 1976; Fellows and Brezonik, 1981). The leached N may also enter deeper groundwater and cause problems which are well documented and known. However the leached N may be largely lost through denitrification as the water enters surface water (Jacobs and Gilliam, 1985; Lowrance et al., 1984; Thomas and Barfield, 1974). The effect of increased leaching is very location specific with regard to potential environmental problems.

SUMMARY AND CONCLUSIONS

We have attempted in this paper to compare the fate of N in conservation tillage or no-till (NT) to the fate of N in conventional tillage. Some very good data are available for some comparisons and a dearth of information for some obvious pathways for losses of N in NT. To reach our conclusions, we used the difference method to estimate the effect of NT on leaching losses in particular.

Volatilization of N as ammonia is a much larger problem with surface applied urea-based fertilizers in NT than in CT. It is impossible to quantitatively estimate the losses of surface applied urea-N as the loss can be very large under some conditions and near zero under others. This poses a significant problem for NT because it is much more difficult to incorporate fertilizers in the conservation tillage systems. Also the fertilizer industry is moving toward urea being a near exclusive source of N fertilizer. An obvious research need is to develop an acceptable method of applying urea in conservation tillage systems without loss of N by volatilization.

Denitrificaiton losses also tend to be greater under NT as compared to CT largely because of the increased moisture in the NT fields. Not all researchers have measured significant losses of N via denitrification in NT fields but when a difference in denitrification rates occurred, the losses were greater in NT. We concluded that average increase in denitrification in conservation tillage as compared to CT might be 2-8 kg N/ha/yr.

There has been a large amount of work quantifying the amount of N which is either mineralized or immobilized under conservation tillage management. It is clear that conversion from CT to any form of conservation tillage will usually result in an increase in organic N present in the soil. Most researchers have found that these increases will continue for many years. However, it has also been shown that conversion of grassland to conservation tillage will generally result in a decrease in soil orgainic N. This very strongly suggests that each system has a steady state level of organic N toward which it is moving. Thus the increases in organic N which occur when going from CT to NT will not continue indefinitely. After a number of years in NT, organic N will neither be a sink for added N nor a source of N for the crop. If this assumption is correct, then fertilizer N – N harvested with crop –(denitrification + volatilization) = N available for contribution to water quality problems. It is our belief that immobilization of N in NT is a short term effect with regard to any water quality problems contributed to by fertilizer N.

Differences in efficiency of fertilizer N between NT and CT have been observed in many experiments although the differences are usually small. In general, the efficiency is higher for the CT at lower N additions and the same or higher for the NT at the higher rates. The differences at the lower N rates are attributed to more immobilization in the NT and the differences at the higher rates are a result of better moisture conditions in the NT. These results probably tend to encourage higher N applications in conservation tillage. Even the higher efficiences of utilization result in significant amounts of N which are not utilized by the crop.

The data on loss of N by surface runoff should be sufficient to convince anyone that NT results in less total N loss via this path as compared to CT. The reduction occurs as a result of a decrease in sediment-associated organic N. The sediment N concentration is usually increased as well as concentration of dissolved N. There is also a small (2 to 4 kg N/ha/yr) increase in the loss of inorganic N. However, the total N lost to surface drainage water is less. This forms the basis for many research reports that conservation tillage results in less potential pollution from use of fertilizers.

Most environmental problems resulting from use of N fertilizer in agricultural production are caused by N which is leached below the rooting zone. There has been very little research on the effect of conservation tillage on amounts of N leached. One study in Kentucky indicated that more added fertilizer N was leached in a killed grass sod than in a CT system. A recent study in Iowa indicated that less native soil N was leached under NT than under CT. These results are both explainable by more water moving through large soil pores under the NT treatments. They also indicate that conservation tillage may increase or decrease leaching in the short term depending upon soil and other factors.

Thus far, we have concluded that: (1) NT increased N loss by volatilization but this loss is likely to be overcome by better fertilizer application methods; (2) denitrification removes slightly more N from NT; (3) immobilization will increase the amount of N present in the soil as organic N but the system will reach a new steady state level of N so this will not be a continual sink for N; (4) N harvested by the crop is not greatly different and (5) there is less loss of N by surface runoff. If the increase in amount of N loss by volatilization and denitrification = the decrease in N in surface runoff, then there would be no difference between the two systems in N lost by leaching. If however, the surface runoff decrease under conservation tillage is greater than the increase in N lost from the system as a gas, then there would be an increase in the quantities of N leached. In either of these situations, conservation tillage has not improved the situation with regard to the greatest environmental problem with regard to N fertilizer use.

In our view, the greatest research need with regard to fate of N under conservation tillage is to obtain the necessary data to prove or disprove the hypothesis that conservation tillage has not improved or may have even increased the environmental problems associated with the use of N fertilizers. Another research need is to develop methods for applying urea in conservation tillage without unacceptable losses of N by volatilization. There is also the continued need for methods to improve fertilizer utilization efficiency.

REFERENCES

Alberts, E. E., W. H. Neibing and W. Moldenhauer. 1981. Transport of sediment nitrogen and phosphorus in runoff through cornstalk residue strips. Soil Sci. Soc. Am. Proc. 45:1177-1184.

Angle, J. S., F. McClung, M. S. McIntosh, P. M. Thomas and D. C. Wolf. 1984. Nutrient losses in runoff from conventional and no-till corn watersheds. J. Environ. Qual. 13:431-435.

Arnott, R. A. and C. R. Clement. 1966. The use of herbicides in alternate husbandry as a substitute for ploughing. Weed Res. 6:142-157.

Aulakh, M. S. and D. A. Rennie. 1984. Transformations of fall-applied nitrogen-15-labelled fertilizers. Soil Sci. Soc. Am. J. 48:1184-1189.

Aulakh, M. S., D. A. Rennie and E. A. Paul. 1984. Gaseous nitrogen losses from soil under zero-till as compared with conventional-till management systems. J. Environ. Qual. 13:130-136.

Baker, J. L. 1985. Conservation tillage: Water quality implications. pp. 217-238. In: F. M. D'Itri (ed.). A System Approach to Conservation Tillage. Lewis Publishers, Inc. Chelsea, MI.

Baker, J. L. and J. M. Laflen. 1982. Effect of corn residue and fertilizer management on soluble nutrient runoff losses. Trans. ASAE 25:344-348.

Baker, J. L. and J. M. Laflen. 1983. Water quality consequences of conservation tillage. J. Soil and Water Cons. 38:186-193.

Baldwin, P. L., W. W. Frye and R. L. Blevins. 1985. Effects of tillage on quality of runoff water. pp. 169-174. In: The Rising Hope of Our Land. Proc. S. Region. No-Till Conf. July 16-17, Griffin, GA.

Bandel, V. A., S. Dzienia and G. Stanford. 1980. Comparison of nitrogen fertilizers for no-till corn. Agron. J. 72:337-341.

Bandel, V. A., S. Dzienia, G. Stanford and J. O. Legg. 1975. N behavior under no-till vs conventional corn culture. I. First year results using unlabeled N fertilizer. Agron. J. 67:782-786.

Barisas, S. G., J. L. Baker, H. P. Johnson and J. M. Laflen. 1978. Effect of tillage systems on runoff losses of nutrients: A rainfall simulation study. Trans. ASAE 21:893-897.

Bauer, A. and A. L. Black. 1981. Soil carbon, nitrogen, and bulk density comparisons in two cropland tillage systems after 25 years and in virgin grassland. Soil Sci. Soc. Am. J. 45:1166-1170.

Beale, O. W., G. B. Nutt and T. C. Peele. 1955. The effect of mulch tillage on runoff, erosion, soil properties and crop yields. Soil Sci. Soc. Am. Proc. 19:244-247.

Blevins, R. L., G. W. Thomas and P. L. Cornelius. 1977. Influence of no-tillage and nitrogen fertilization on certain soil properties after 5 years of continuous corn. Agron. J. 69:383-386.

Blevins, R. L., G. W. Thomas, M. S. Smith, W. W. Frye and P. L. Cornelius. 1983. Changes in soil properties after 10 years continuous no-tilled and conventionally tilled corn. Soil Tillage Res. 3:135-146.

Broder, M. W., J. W. Doran, G. A. Peterson and C. R. Fenster. 1984. Fallow tillage influence on spring populations of soil nitrifiers, denitrifiers, and available nitrogen. Soil Sci. Soc. Am. J. 48:1060-1067.

Buhtz, E., O. Bosse, R. Herzog and U. Waldschmidt. 1970. Ergebnisse zur Rationaliasierung der Gundbodenbearbeitung. Albrecht-Thaer-Arch. 14:795-812.

Carter, M. R. and D. A. Rennie. 1984. Crop utilization of placed and broadcast 15-N-urea fertilizer under zero and conventional tillage. Can. J. Soil Sci. 64:563-570.

Chichester, F. W. 1976. The impact of fertilizer use and crop management on nitrogen content of subsurface water draining from upland agricultural watersheds. J. Environ. Qual. 5:413-416.

Doren, J. W. 1980. Soil microbial and biochemical changes associated with reduced tillage. Soil Sci. Soc. Am. J. 44:765-771.

Doran J. W. and J. F. Power. 1983. The effects of tillage on the nitrogen cycle in corn and wheat production. pp. 441-455. In: R. R. Lowrance, R. L. Todd, L. E. Asmussen and R. A. Leonard (eds.). Nutrient cycling in agricultural ecosystems. Georgia Agr. Spec. Pub. 23.

Dowdell, R. J., R. Crees and R. Q. Cannell. 1983. A field study of contrasting methods of cultivation on soil nitrate content during autumn, winter, and spring. J. Soil Sci. 34:367-379.

Dowdell, R. J. and R. Q. Cannell. 1975. Effect of ploughing and direct drilling on soil nitrate content. J. Soil Sci. 26:53-61.

El-Haris, M. K., V. L. Cochran, L. F. Elliott and D. F. Bezdicek. 1983. Effect of tillage, cropping, and fertilizer management on soil nitrogen mineralizaiton potential. Soil Sci. Soc. Am. J. 47:1157-1161.

Fellows, C. R. and P. L. Brezonik. 1981. Fertilizer flux into two Florida lakes via seepage. J. Environ. Qual. 10:174-177.

Ferrer, M. B., R. N. Gallaher and R. G. Volk. 1984. Soil nitrogen and organic matter changes as affected by tillage after six years of corn. pp. 189-192. In: J. T. Touchton and R. E. Stevenson (eds.). Proc. of the Seventh annual Southeast no-tillage systems conf. Auburn Agr. Expt. Sta.

Fleige, H. and K. Baeumer. 1974. Effect of zero-tillage on organic carbon and total nitrogen content, and their distribution in different N-fractions in loessial soils. Agro-Ecosystems 1:19-29.

Free, G. R. 1970. Minimum tillage for corn production. New York (Cornell) Agr. Exp. Stn. Bul. 1030.

Gilliam, J. W. 1985. Fertilizer. pp. 189-192. In: McGraw-Hill Yearbook of Science and Technology. McGraw-Hill, New York.

Gilliam, J. W., T. J. Logan and F. E. Broadbent. 1985. Fertilizer use in relation to the environment. pp. 561-588. In: O. P. Engelstad (ed.). Fertilizer Use and Technology. Am. Soc. Agron., Madison, WI.

Gold, A. J. and T. L. Loudon. 1982. Nutrient, sediment and herbicide losses in tile drainage under conservation tillage. Paper No. 82-2549. ASAE, St. Joseph, MI.

Goss, M. J., K. A. Hawse and W. Harris. 1978. Effects of cultivation on soil water retention and water use by cereals in clay soils. J. Soil Sci. 29:475-488.

Groffman, P. M. 1985. Nitrification and denitrification in conventional and no-tillage soils. Soil Sci. Soc. Am. J. 49:329-334.

Harrold, L. L., G. B. Triplett, Jr. and W. M. Edwards. 1970. No-tillage corn-characteristics of the system. Agr. Eng. 51:128-131.

House, G. J., B. R. Stinner, D. A. Crossley, Jr. and E. P. Odum. 1984. Nitrogen cycling in conventional no-tillage agro-ecosystems: Analysis of pathways and processes. J. Appl. Ecol. 21:991-1012.

House, G. J., B. R. Stinner, D. A. Crossley, Jr. E. P. Odum and G. W. Langdale. 1984. Nitrogen cycling in conventional and no-tillage agroecosystems in the Southern Piedmont. J. Soil Water Cons. 39:194-200.

Jacobs, T. J. and J. W. Gilliam. 1985. Riparian losses of nitrate from agricultural waters. J. Environ. Qual. 14:472-478.

Johnson, H. P., J. L. Baker, W. D. Shrader and J. M. Laflen. 1979. Tillage system effects on sediment and nutrients in runoff from small watersheds. Trans. ASAE 22:1110-1114.

Kahnt, G. 1971. NPK- and C-Veranderungen auf 3 Bodentypen nach 5 Jahren pfluglosem Ackerbau. Landirtsch. Forsch. S. H., 26/1:273-280.

Kanwar, R. S., J. L. Baker and J. M. Laflen. 1985. Nitrate movement through the soil profile in relation to tillage system and fertilization application method. Trans. ASAE 28:1802-1807.

Kitur, B. K., M. S. Smith, R. L. Blevins and W. W. Frye. 1984. Fate of N-depleted ammonium nitrate applied to no-tillage and conventional tillage corn. Agron. J. 76:240-242.

Laflen, J. M., J. L. Baker, R. O. Hartwig, W. F. Buchele and H. P. Johnson. 1978. Soil and water loss from conservation tillage systems. Trans. ASAE 21:881-885.

Laflen, J. M. and T. S. Colvin. 1981. Effect of crop residue on soil loss from continuous row cropping. Trans. ASAE 24:605-609.

Laflen, J. M. and M. A. Tabatabai. 1984. Nitrogen and phosphorus losses from corn-soybean rotations as affected by tillage practices. Jour. ASAE 27:58-63.

Lal, R. 1976. No-tillage effects on soil properties under different crops in western Nigeria. Soil Sci. Soc. Am. J. 40:762-768.

Linn, D. M. and J. W. Doran. 1984a. Aerobic and anaerobic microbial populations in no-till and plowed soils. Soil Sci. Soc. Am. J. 48:794-799.

Linn, D. M. and J. W. Doran. 1984b. Effect of water-filled pore space on carbon dioxide and nitrous oxide production in tilled and nontilled soils. Soil Sci. Soc. Am. J. 48:1267-1272.

Lowrance, R. R., R. L. Todd and L. E. Asmussen. 1984. Nutrient cycling in an agricultural watershed: I. Phreatic movement. J. Environ. Qual. 13:22-27.

McDowell, L. L. and K. C. McGregor. 1980. Nitrogen and phosphorus losses in runoff from no-till soybeans. Trans., ASAE 23:643-648.

Mengel, D. B., D. W. Nelson and D. M. Hefer. 1982. Placement of nitrogen fertilizer for no-till and conventional till corn. Agron J. 74:515-518.

Moschler, W. W., G. M. Shear, D. C. Martens, G. D. Jones and R. R. Wilmouth. 1972. Comparative yield and fertilizer efficiency of notillage and conventionally tilled corn. Agron. J. 64:229-231.

Onstad, C. A. 1972. Soil and water losses as affected by tillage practices. Trans., ASAE 15:287-289.

Ortiz, R. A. and R. N. Gallaher. 1984. Organic matter and nitrogen in an ultisol as affected by cropping and tillage system after seven years. In: J. T. Touchton and R. E. Stevenson (eds.). Proc. of the Seventh Annual Southeast No-tillage Systems Conf. Auburn Agr. Expt. Sta.

Powlson, D. S. and D. S. Jenkinson. 1981. A comparison of the organic matter, biomass, adenosine triphosphate and mineralizable nitrogen contents of ploughed and direct-drilled soils. J. Agric. Sci. UK 97:713-721.

Rice, C. W. and M. S. Smith. 1982. Denitrification in no-till and plowed soils. Soil Sci. Soc. Am. J. 46:1168-1173.

Rice, C. W. and M. S. Smith. 1983. Nitrification of fertilizer and mineralized ammonium in no-till and plowed soil. Soil Sci. Sci. Am. J. 47:1125-1129.

Romkens, M. J. M., D. W. Nelson and J. B. Mannering. 1973. Nitrogen and phosphorus composition of surface runoff as affected by tillage method. J. Environ. Qual. 2:292-295.

Siemens, J. C. and W. R. Oschwald. 1978. Corn-soybeans tillage systems: Erosion control, effects on crop production cost. Trans., ASAE 21:293-302.

Stanford, G., O. L. Bennett and J. F. Power. 1973. Conservation tillage practices and nutrient availability. pp. 54-62. In: Conservation Tillage. Proc. Nat. Conserv. Tillage Conf. (28-30 March 1973, Des Moines, Iowa) Soil Conserv. Soc. Am.

Stinner, B. R., G. D. Hoyt and R. L. Todd. 1983. Changes in soil chemical properties following a 12-year fallow: A 2-year comparison of conventional tillage and no-tillage agro-ecosystems. Soil Tillage Res. 3:277-290.

Thomas, G. W. and B. J. Barfield. 1974. The unreliability of tile effluent for monitoring subsurface nitrate-nitrogen losses from soils. J. Environ. Qual. 3:183-185.

Thomas, G. W., R. L. Blevins, R. E. Phillips and M. A. McMahan. 1973. Effect of a killed sod mulch on nitrate movement and corn yield. Agron. J. 63:736-739.

Thomas, G. W., K. L. Wells and L. Murdock. 1981. Fertilization and Liming. pp. 43-54. In: R. E. Phillips et al. (eds.). No Tillage Research: Research Reports and Reviews. Univ. KY, Lexington, KY.

Timmons, D. R., R. F. Holt and J. J. Latterell. 1970. Leaching of crop residues as a source of nutrients in surface runoff water. Water Resources Res. 6:367-375.

Touchton, J. T. and W. L. Hargrove. 1982. Nitrogen sources and method of application for no-tillage corn production. Agron. J. 74:823-826.

Tyler, D. D. and G. W. Thomas. 1979. Lysimeter measurement of nitrate and chloride losses from conventional and no-tillage corn. J. Environ. Qual. 6:63-66.

Wells, K. L. 1984. Nitrogen management in the no-till system. pp. 535-550. In: R. D. Hauck (ed.). Nitrogen in Crop Production. Am. Soc. Agron., Madison, WI.

Wendt, R. C. and R. E. Burwell. 1985. Runoff and soil losses for conventional, reduced and no-till corn. J. Soil and Water Cons. 40:450-454.

CHAPTER 14

EFFECT OF CONSERVATION TILLAGE ON PROCESSES AFFECTING NITROGEN MANAGEMENT

J. S. Schepers
USDA-ARS, University of Nebraska, Lincoln, Nebraska

INTRODUCTION

Conservation tillage systems modify a number of chemical, biological and physical processes in soils relative to more traditional tillage practices that disturb the soil and leave little residue cover. The acceptance of conservation tillage systems by producers is frequently determined by the relative economic advantage of such systems. However, a number of factors and processes are involved that may not be evaluated except in terms of final yield. Gilliam and Hoyt (Chapter 13) present an excellent compilation of conservation tillage research which should prove to be a valuable reference and significantly contribute to an expression of the dynamics of nitrogen (N) associated with tillage systems. Little more can be contributed in terms of a literature review that would further elucidate the effects of tillage systems on N volatilization, denitrification, mineralization, immobilization, runoff, and nitrate leaching. The approach of this response paper will be to examine some of the interactions between the chemical, biological, and physical processes involved in tillage systems in an attempt to systematically evaluate the potential effect on groundwater quality.

NITRATE LEACHING

The physical process of nitrate leaching requires sufficient soil water for percolation and availability of nitrate ions in close proximity to the moving water. Soils with little structure transmit percolating water very rapidly and readily contribute any available nitrate to the solution. Finer

Effects of Conservation Tillage on Groundwater Quality: Nitrates and Pesticides, Terry J. Logan et al., eds.

textured soils tend to promote better structure, especially in the root zone where organic matter is more prevalent, and reduce the percolation rate. However, fine textured soils do not always have low infiltration rates. While organic matter is noted for its binding and cementing characteristics, changes in soil-water content caused by climate and evaporation can result in cracking of highly structured soils, where a large portion of any percolating water can be transmitted through the larger pores. Root channels and cavities left by earthworm and other faunal activity can also result in macropores. Under such conditions, the smaller pores are less accessible to the percolating water and may not readily contribute to nitrate leaching. Soils containing predominantly montmorillonite and vermiculite type clays tend to promote better aggregation and therefore the micropore-macropore contrasts may become more important in terms of nitrate leaching. It should be noted that even in finer textured soils, good soil structure and aggregation usually does not prevail below the root zone and in many cases below the surface 0.6m.

Soil-water content flucations are usually greater in the root zone than in the vadose zone because of plant growth and evaporation. Therefore, climatic factors and processes within the root zone are the most likely to influence groundwater quality. Physical, chemical, and biological processes in the vadose zone of sandy soils have received limited attention in the past because nitrates may be very transitory under these conditions and sample variations are expected to be large. Infiltration and percolation are slower in finer textured soils and nitrates may require decades to reach groundwater, which permits monitoring nitrate leaching over time. Schuman et al., (1975) were able to document nitrate leaching under conservation tillage in western Iowa, and concluded nitrate leaching was largely influenced by the amount of fertilizer N applied. Muir et al., (1973) came to a similar conclusion when they found groundwater nitrate-N concentrations in Nebraska were positively correlated with fertilizer N applications. They also found groundwater nitrate-N concentrations were positively correlated with irrigation well density but negatively correlated with soil clay content (Table 1). Such relationships are not suprising, but they take time to develop and as indicated by N tracer studies, nitrate other than that from the immediately preceding N fertilizer may also be involved in the leaching process. Other potential sources of nitrate include residual soil N (nitrate left over from the previous crop) and mineralized N. While conservation tillage may alter the N form over a period of months to a few years, the total amount of root zone N may not vary greatly with tillage system.

Table 1. Correlations between groundwater nitrate-N concentrations and site characteristics in Nebraska (Muir et al., 1973).

Correlation of groundwater NO_3-N concentration with:	r value
% clay in soil	-0.49**
Irrigation well density	0.43**
Total fertilizer used	0.28*
Irrigation well depth	-0.28*
Water pH	-0.23*
Livestock density	0.18*
% land cultivated	-0.07
Human population	-0.06
% land with legume	-0.04

Variations in fertilizer N use efficiency ranging from 20 to 80% have been reported using tagged-N, however values of 30 to 60% are more common (Power, 1981). Even with such well controlled studies, it is hard to account for all of the applied N, and losses are usually attributed to leaching, volatilization, or denitrification. Following the tagged-N remaining in the soil organic fraction can contribute to our understanding of N cycling under different tillage systems. Many of the previous tillage comparison studies reach different conclusions as noted by Gilliam and Hoyt (Chapter 13). These inconsistencies may be disconcerting to some, but not surprising, considering the range in soils and climatic conditions influencing the various processes involved.

DYNAMIC PROCESSES

Many processes contribute to the nitrate leaching potential of a soil with associated tillage systems. Simply examining mineralization, nitrification and denitrification processes may not adequately address the complexitites of tillage systems and therefore generalities should be carefully examined for unique features of the soil, location and climate. Inconsistencies between studies comparing conventional tillage and no-tillage can frequently be accounted for by examining the studies on a site specific basis. By doing so, the chemical, physical, and biological processes that interact with climate, soil and cropping systems can be manifested.

It is difficult to discuss how tillage systems influence indivudual processes because each process in turn responds to the biological, chemical, and physical characteristics of that environmental niche. It is certain, however, that biological processes respond to the chemical and physical conditions of their environment in a matter of hours or days. Soil biological activity decreases rapidly with depth and the effect of tillage on microbial processes is largely expressed in the tillage zone, with the exception being soil compaction caused by tillage which alters the soil physical conditions. Since biological processes play a major role in N cycling and potential nitrate leaching, it is perhaps appropriate to examine how biological processes in general respond to climate and soil physical conditions. At this point it may not be important to make a distinction between tillage systems, but rather what tillage operations or lack of them do to the soil physical environment.

Any tillage operation that changes soil bulk density in turn modifies soil pore size distribution, water holding capacity, infiltration rate, and soil aeration. As any or all of these changes occur, the suitability of that environment for a specific group of organisms changes. The net result may be an increase or decrease in activity for each group of organisms. In terms of N cycling, these changes could stimulate aerobic processes at the expense of the anaerobic processes, or vice-versa. Since each soil type and cropping system responds differently to tillage, what may be a desirable tillage system in one location may be complete failure in another. An example is changing from conventional tillage to no-tillage in western Nebrasks (Alliance silt loam) and northern Illinois (Blount silt loam). On the Alliance soil (well drained), no-tillage results in improved water storage, and greater potential yield (presumedly more potential N uptake). In contrast, introducing no-tillage to the Blount soil (imperfectly drained) results in a cooler, partially anaerobic environment where denitrification losses can be large.

In many locations, no-tillage is not an option in the minds of many producers because of problems encountered with subsequent cultural practices. Many producers on fine textured soils in Illinois, Iowa, and Minnesota feel that they must plow or deep till the soil in the fall to improve drainage and aid in warming the soil for spring planting. Producers using furrow irrigation or ridge tillage also find no-tillage unacceptable because of increases in fertilizer N costs if anhydrous ammonia is not used. For these reasons, many producers have adopted various degrees of conservation tillage to fit their cropping systems, soil, and climate. In so doing, these producers have, perhaps inadvertently, integrated a number of biological, chemical and physical properties that result in an acceptable production system. Such tillage systems, however, may not be in the best interest of groundwater quality and unless an integrated approach is taken, it will be difficult to describe the success or failure of conservation tillage systems.

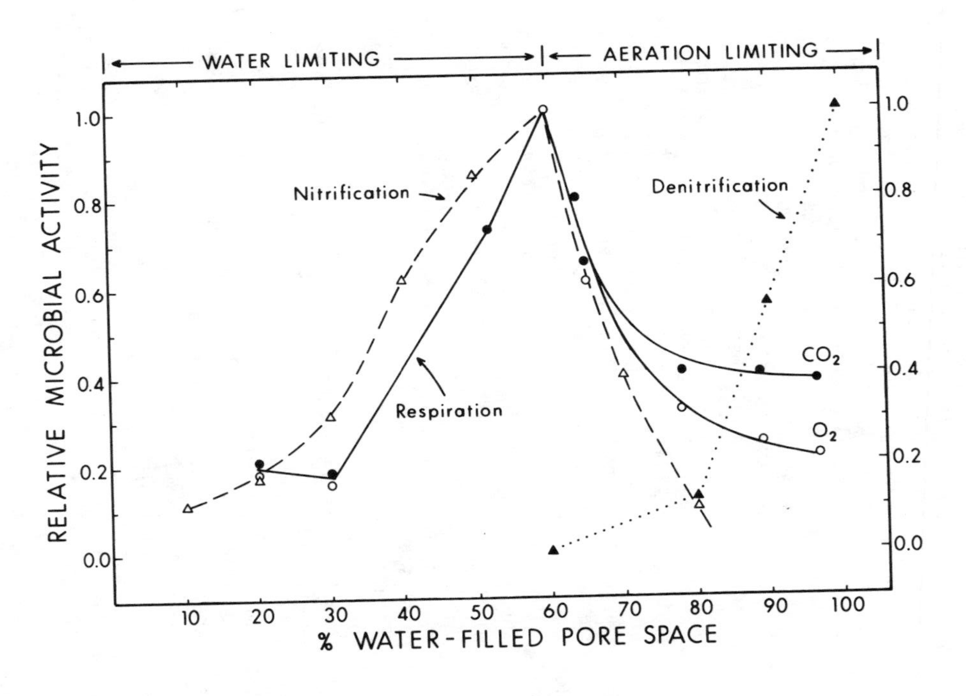

Figure 1. The relationship between water-filled pore space and relative amount of microbial nitrification, denitrification, respiration (O_2 uptake), and CO_2 production (Linn and Doran, 1984).

INTERGRATED ENVIRONMENTAL CONDITIONS

The dynamics of soil biological processes are known to be a function of temperature, water status, aeration, pH and substrate availability. All of these soil properties can be influenced by tillage and, for that reason, each must be considered when evaluating a tillage system. Fortunately, N mineralization is insensitive to pH between values of 6.2 and 9.0; however nitrification only proceeds at its maximum rate between pH 7.0 and 7.4 (Reddy et al., 1979). For this reason, ammonium-N may temporarily dominate the inorganic N pool of acidic surface soils, especially under no-tillage where surface applied N and mineralization of surface residues have acidified the soil. Ammonium-N is less susceptible to leaching than nitrate-N, and hydroponic studies with wheat (Dr. R. H. Hageman, personal communication of unpublished data) and field studies with corn (Pan et al., 1984) indicate plants prefer a combination of nitrate-and ammonium-N sources, which may be a positive attribute of no-tillage systems.

Soil water status and aeration can be integrated into one soil property when combined with bulk density to give an expression referred to as relative water saturation or percent water-filled pore space (Greaves and Carter, 1920; Linn and Doran, 1984). The merit of this calculated parameter is that it only requires determination of water content and bulk density, but in terms of microbial activity, it can be used to indicate the relative importance or likelihood of net mineralization and denitrification. For most soils, it appears that a value of 60% water-filled pore space is near optimum for aerobic activity, while denitrification may not become important until 80% water-filled pore space or more is reached (Figure 1). Very sandy soils appear to be the exception to the above guidelines, where 90% water-filled porespace may be required to maximize mineralization (Watts, 1975). these conditions could only be attained in a tumbling incubation system and the likelihood of sandy soils remaining at or near saturated conditions is very small except where drainage is a problem, and in fact well drained sandy soils may seldom reach the required water content for optimum mineralization.

The relative water saturation concept can also be used to evaluate the impact of tillage on microbial processes. Precipitation or irrigation will reduce soil aeration status, and depending on the soil type and tillage system, may result in more than 60% of the pores filled with water and a greater potential for denitrification (Figure 2). Tillage of wet soils may aerate some portions of the soil and thereby increase mineralization, but other portions may become compacted and promote denitrification. Wheel traffic areas and tillage pans are examples that can be predicted using this concept. Recent versions of a nitrogen-tillage-residue-management simulation model (NTRM) incorporate this concept into biological process subroutines (Clay et al., 1985; Shaffer, 1985). The time is

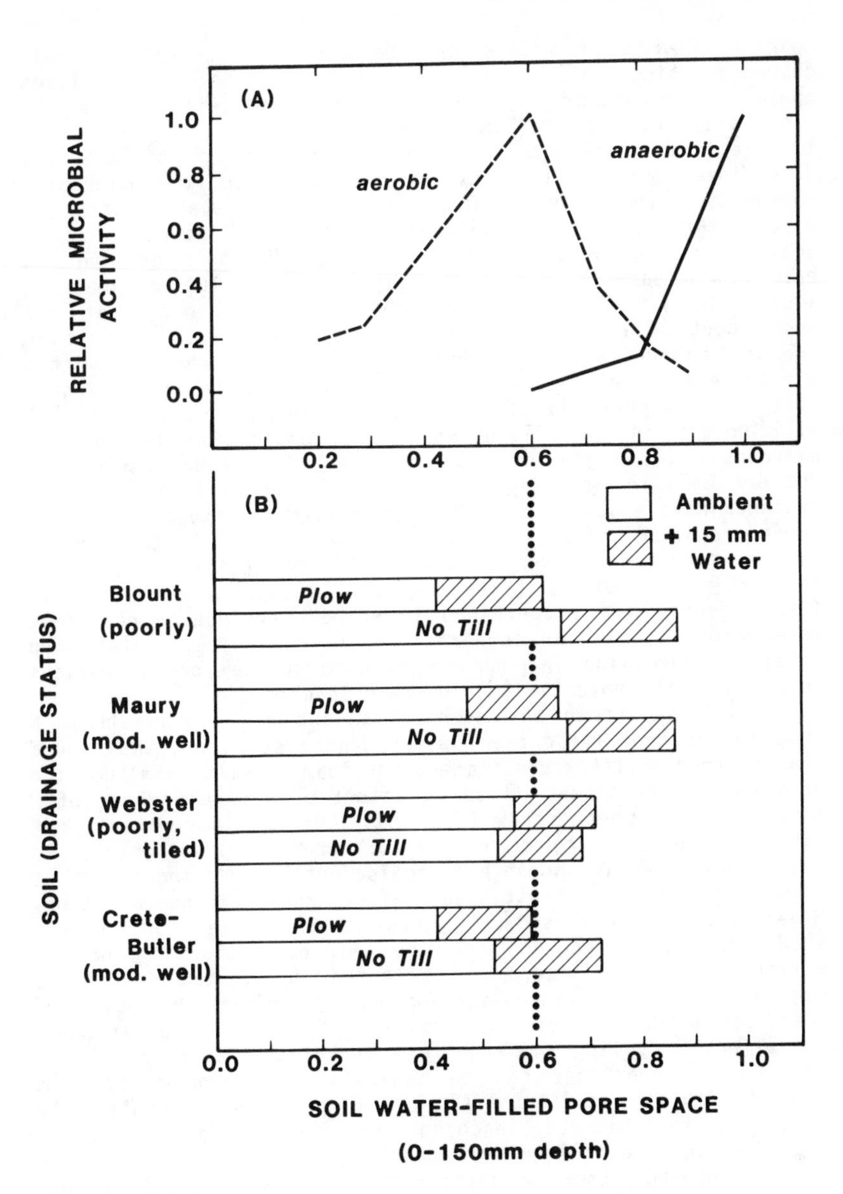

Figure 2. Effect of adding water to several soils under different tillage systems relative to the suggested value of 60% water-filled pore space (Mielke et al., 1986).

nearing when such models can be used to incorporate soil characteristics and climatic factors to describe tillage operations or other cultural practices that will create the disirable soil microbial environment.

Residue placement is a direct effect of tillage and determines the availability of substrate to the microbial community. The combined effect of substrate availability and placement relative to a soil environment conductive to microbial activity can greatly effect N cycling. Nitrogen cycling can have a number of crop availability and nitrate leaching implications. Substrates with carbon-to-nitrogen ratios of less than about 30:1 (greater than 1.2 to 1.5% N) will result in "normal" mineralization and release mineral N to the soil within a relatively short time (Doran et al., 1977; Smith and Peterson, 1982). Incorporation of residues containing less than 1.2% N will result in immobilization and a competition between the microbes and the growing crop for soil mineral N. As pointed out by Gilliam and Hoyt (Chapter 13), no-tillage involving residues have a large C:N ratio can result in immobilization of surface applied N fertilizer and poor utilization by the crop in semi-arid regions (Power et al., 1986), but better utilization in semi-humid regions (Kitur et al., 1984). Power et al., (1986) found that N contained in soybean residues mineralized more readily than N in corn residue, which indicates that crop rotations involving legumes may require special considerations to minimize the potential for nitrate leaching.

Little is known about crop utilization of mineralized N from recent crop residues, however Wagger et al., (1985) found that neither soil texture (fine sandy loam or silt loam) nor the addition of fertilizer N had an effect on the net amount of N mineralized in wheat straw (C:N of ∿ 100:1). In contrast, net mineralization of sorghum residue (C:N of ∿ 30:1) was greater in the silt loam soil, and in both soils fertilizer N increased net mineralization the first year after residue incorporation. Mineralization in subsequent years of the above study is not known, but given enough time, net mineralization may not be affected by fertilizer N as noted by Sorensen (1982). Regardless of net mineralization, it is clear that type of residue, residue placement, fertilizer N, and soil environment can effect N cycling and therefore N availability to the crop. If cultural practices can be manipulated to synchronize the availability of soil mineral N and crop requirements, the potential for nitrate leaching should be greatly reduced, regardless of the tillage system.

Evaluating the potential for nitrate leaching and subsequent groundwater pollution under no-tillage production systems is difficult. As noted by Meisinger (1984), "Since N is a mobile and dynamic nutrient, it is the one element where a comprehensive approach is not only suggested - it is required". Producer acceptance of certain conservation tillage practices may result for many of the same reasons as they are considered environmentally sound, because processes that result in N losses also represent an economic loss. Expanded knowledge dealing with macropore water flow and subsequent leaching of nitrate and pesticides will be required before the impact of conservation tillage systems is completely evaluated.

REFERENCES

Clay, D. E., J. A. E. Molina, C. E. Clapp, and D. R. Linden. 1985. Nitrogen-tillage-residue management: II. Calibration of potential rate of nitrification by model simulation. Soil Sci. Soc. Am. J. 49:322-325.

Doran, J. W., J. R. Ellis, and T. M. McCalla. 1977. Microbial concerns when wastes are land applied. p. 343-362. In Proc. 1976 Cornell Agric. Waste Mgt. Conf., Ann Arbor Science.

Greaves, J. E., and E. G. Carter. 1920. Influence of moisture on the bacterial activity of the soil. Soil Sci. 10:361-387.

Kitur, B. K., M. S. Smith, R. L. Blevins, and W. W. Frye. 1984. Fate of ^{15}N-depleted ammonium nitrate applied to no-tillage and conventional tillage corn. Agron. J. 76:240-242.

Linn, D. M., and J. W. Doran. 1984. Effect of water-filled pore space on carbon dioxide and nitrous oxide production in tilled and nontilled soils. Soil Sci. Soc. Am. J. 48:1267-1272.

Meisinger, J. J. 1984. Evaluating plant-available nitrogen in soil crop systems. Roland D. Hauck (ed.) In Nitrogen in Crop Production. p. 391-416.

Mielke, L. N., J. W. Doran, and K. A. Richards. 1986. Physical environment near the surface of plowed and no-till soils. Soil and Tillage Res. 7:355-366.

Muir, J., E. C. Seim, and R. A. Olson. 1973. A study of factors influencing the nitrogen and phosphorus contents of Nebraska water. J. Environ. Qual. 2:466-469.

Pan, W. L., E. J. Kamprath, R. H. Moll, and W. A. Jackson. 1984. Prolificacy in corn: its effects on nitrate and ammonium uptake and utilazaiton. Soil Soc. Am. J. 48:1101-1106.

Power, J. F. 1981. Nitrogen in the cultivated Ecosystem. F. E. Clark and T. Rosswall (eds) In Terrestrial Nitrogen Cycles, Ecol. Bull. (Stockholm) 35:529-546.

Power, J. F., J. W. Doran, and W. W. Wilhelm. 1986. Uptake of nitrogen from soil, fertilizer, and crop residues by no-till corn and soybean. Soil Sci. Soc. Am. J. 50:137-142.

Reddy, K. R., R. Khaleel, M. R. Overcash, and P. W. Westerman. 1979. A nonpoint source model for land career receiving annual wastes: I. Mineralization of organic nitrogen. Trans. Am. Soc. Agric. Engr. 22:863-872.

Schuman, G. E., T. M. McCalla, K. E. Saxton, and H. T. Knox. 1975. Nitrate movement and its distribution in the soil profile of differentially fertilized corn watersheds. Soil Sci. Soc. Am. Proc. 39:1192-1197.

Shaffer, Marvin J. 1985. Simulation model for soil erosion-productivity relationships. J. Environ. Qual. 14:144-150.

Smith, J. H. and J. R. Peterson. 1983. Recyclying of nitrogen through land application of agricultrual, food processing, and municipal wastes. p. 791-831. In Nitrogen in Agricultural Soils, Agronomy Mono. No. 22.

Sorensen, L. H. 1982. Mineralization of organically bound nitrogen in soil as influenced by plant growth and fertilization. Plant and Soil 65:51-61.

Wagger, M. G., D. E. Kissel, and S. J. Smith. 1985. Mineralization of nitrogen from nitrogen- 15 labeled crop residues under field conditions. Soil Sci. Sco. Am. J. 49:1220-1226.

Watts, D. J. 1975. A soil-water-nitrogen-plant model for irrigated corn on coarse textured soils. Ph.D. diss. Utah State Univ., Logan (Diss. Abst. 76-06253).

SECTION IV

SELECTED TOPICS ON CONSERVATION TILLAGE SYSTEMS

CHAPTER 15

MANURE MANAGEMENT WITH CONSERVATION TILLAGE

M. F. Walter, T. L. Richard and P. D. Robillard,
Cornell University, Ithaca, New York

R. Muck,
University of Wisconsin, Madison, Wisconsin

INTRODUCTION

The connection between soil and water resources is so intimate that any change in one usually affects the other. It was assumed for many years that soil conservation practices not only protected the soil but enhanced water quality as well. This assumption was first seriously questioned in the early 1970s as environmentalists began to look for "best management practices" (BMPs) for control of agricultural nonpoint source pollution (NPS). The general notion that soil and water conservation practices (SWCPs) were good for water quality was neither specific nor substantiated enough for blanket acceptance of SWCPs as BMPs, thus a number of questions have been raised. Which soil conservation practices are candidate BMPs for which water pollutants? How specifically do the various SWCPs affect the fate, and particularly the transport of different potential water contaminants? Might not some SWCPs result in even greater use of chemicals that are potential water quality contaminants? In general, we found that when a practice is applied with a specific objective in mind, (e.g., conservation tillage for soil erosion control or reduced fertilizer use to prevent groundwater pollution) changes occur which affect the total system - including physical, economic and even social components - in complex ways. As an old axiom states, "There are many simple answers to complex questions, but most of them are wrong."

Effects of Conservation Tillage on Groundwater Quality: Nitrates and Pesticides, Terry J. Logan et al., eds.

Many of the issues related to protecting environmental quality become even more complex when we consider resources other than just soil and water. Two of these are labor and energy, and these are tied very closely to a third, livestock manure. The interaction of these resources mandates that we address single purpose goals from a system approach. We must seek to optimize the use and protection of all resources together, not one at a time.

CONFLICTING OBJECTIVES

Conservation tillage maintains crop residue at the surface as a means to reduce erosion and protect soil resources. An important goal in land application of manure is to incorporate the manure in the soil to conserve the manurial nutrients and reduce surface water contamination. Nearly every recommendation for surface application of manure on cropland suggests incorporating the manure as soon as possible following land spreading (LWFH, 1985; Gilbertson, et al. 1979: Cornell Recommends, 1985). It would appear that the attempt to maintain residues on the surface through use of conservation tillage might be in conflict with that of incorporating manure. However, this may not be true in all cases as we will endeavor to show.

Much of the discussion in this paper deals with our experience with dairy farms. In general, we feel our observations are applicable to other types of livestock farming but we recognize that each situation is unique. Differences in solids and nutrient content among manures from different livestock will affect handling, storage, and application technologies, and some of our comments about nutrient management and water quality protection will not apply in all cases. Nonetheless, the discussion should apply in a general way to most livestock systems, with modifications where appropriate. For purposes of this paper, we are focusing our dicussion on manurial nitrogen and phosphorus and not other manurial water contaminants.

WHO CARES AND WHY

In discussing the role of manure management in a conservation tillage program, it is useful to distinguish between "on-farm" needs and objectives and downstream effects. These two areas are represented by different groups of people with different priorities, and the solution to a problem in one area can have negative effects on the other. Sometimes, however, a particular practice can have benefits for both groups. For example, downstream concerns with sedimentation from soil erosion and farmer attempts to minimize energy and labor costs are both benefited by conservation tillage practices.

We believe that conservation tillage can be an attractive alternative for farmers because it is less labor and energy intensive and therefore potentially more profitable. In addition, is an extremely effective soil erosion control practice. Conservation tillage can in some cases be an effective measure for water quality control and on-farm fuel conservation. Because of their multiple benefits, conservation tillage has been widely supported by local, state, regional, national and even international programs. These programs have included technical assistance, education, cost-sharing and generally have been tailored to crop production. However, the somewhat different concerns of livestock farmers have been largely overlooked in these important conservation tillage programs.

In the United States, conservation tillage has not been as rapidly adopted in the Northeast as it has been in the Midwest, particularly among livestock farmers. The innovative pilot tillage programs in Oswego and Wayne Counties (New York) have documented several agronomic problems linked to conservation tillage on manure spread fields. Low seed germination, decreased effectiveness of herbicide treatments and slug problems were associated with direct seeding into fields which had received heavy surface manure applications. The poor performance of no-till on manure spread fields caused the pilot program to avoid these fields or to recommend very low rates of surface manure application before no-till planting. Even in those cases where low rate applications were attempted, equipment and weather conditions limited the ability of the operator to spread a low uniform rate. These experiences point out the need for conservation tillage practices to be tailored to livestock cropping systems before they can be effective and acceptable to farmers.

Livestock manure that is collectible is concentrated geographically. Gilbertson, et al. (1979) shows a map of areas where manure from livestock and poultry can be collected and spread economically. As shown in Figure 1, there appear to be three geographic concentrations of special interest including: (1) New York, Pennsylvania and Vermont; (2) Wisconsin, Iowa, southern Minnesota, northern Illinois, eastern South Dakota and eastern Nebraska; and (3) southern California and New Mexico. The concern for identifying cost effective manure management-conservation tillage systems is probably greater in these areas than in other parts of the country.

Particularly in these regions where large areas are subject to surface application of manure, downstream users, represented by water quality regulation agencies, have a legitimate concern with protecting surface and groundwater from excessive nutrient pollution. BMPs as they are presently developed are an attempt to promote techniques which protect those water resources, but our understanding of the interactions between techniques within a system is not yet complete. Perhaps due to the historically limited use of conservation tillage by livestock farmers, we are only beginning to learn what the water quality effects are of manure application with limited tillage.

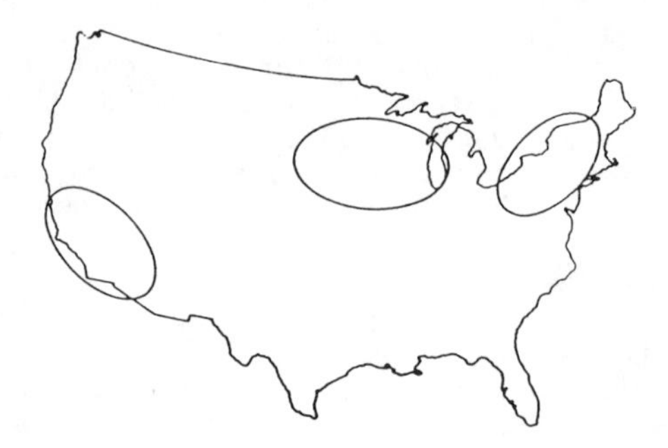

Figure 1. Regional areas of concentration of livestock and poultry which are economically collectible in the continental United States, 1974 (Adapted from Gilbertson, et al. 1979).

Similarly, not enough attention has been paid to the on-farm benefits and costs due to different methods of manure management under a reduced tillage program. To get a better sense of what those tradeoffs might be, it seems appropriate to discuss the use of manures in the farming system from a resource efficiency perspective.

MANAGEMENT OF FARM RESOURCES: ENERGY, LABOR AND NUTRIENTS

Conservation tillage has been accepted by farmers largely because increased chemical costs have been more than offset by reduced fuel and labor costs, while still producing an economic crop yield. Different manure application methods each have their own tradeoffs among these resources. As discussed earlier, maximum plant utilization of manurial nutrients depends on rapid and complete incorporation into the soil. Surface spreading along with tillage necessitates at least two trips over the land receiving manure. In general, greater manurial nutrient recovery requires increased energy and labor inputs by the farmer. Over the past few decades we have seen the relative costs of these inputs fluctuate significantly, and overall farm management has changed as a result.

Efficient utilization of the farm labor resource has been an important factor in farmer selection of both tillage and manure management programs. One study found that the labor required for chisel tillage was only 75% of that for moldboard plowing (IDES, 1986). Labor savings make this practice attractive to livestock farmers as well as those who raise only cash crops. Simiarly, "good manure management" is often sold on the basis of utilizing the manurial nutrients or even as a means for water quality control, but farmers often see it as a convenience as a means to improve their overall farm management. For example, inclusion of manure storage in a waste handling system greatly improves the farmer's flexibility in land application options, and storage can save the farmer's time. The IDES (1986) study found that 421 hours were required annually to spread manure daily from the study farm, but this was reduced to only 190 hours when storage was used.

In the early 1980s a great deal of attention was focused on conservation of farm energy resources. Both conservation tillage and good manure management including incorporation are viewed as important energy conservation measures. Conservation tillage is a relatively low energy consuming tillage practice requiring only about half of the energy of moldboard plowing (IDES, 1984). Manurial nutrients can be used as a substitute for chemical fertilizer which is one of the highest energy requirements for crop production. Gunkel (1974) lists nitrogen fertilizer as the single greatest energy component of corn production in the United States. Clearly, there are energy saving opportunities in a farming strategy which include both of these practices.

The scarcity or abundance of livestock manures largely dictates the efficiency with which farm nutrient resources must be handled. Farms with large livestock operations and little land on which to spread manure treat the manurial nutrients as a waste which must be disposed. Conversely, farms with more cropland often find that the energy and labor required for manure spreading are recovered in increased yields, improved soil tilth, and decreased expense for petrochemical fertilizers. Because the farm nutrient balance has such a critical impact on manure management strategies, this resource is more fully described in the following section.

ANIMAL MANURE AS A RESOURCE/WASTE

For a variety of reasons, mostly economic, farmers in the United States have moved toward monocultures and there has been concentration of certain types of agriculture in geographic regions. This has led to a breakdown in the natural cycle that occurred on polycultural livestock-cropping farms in the past. For example, Figure 2 shows the basic nutrient cycle for a livestock-cropping system. In general, the livestock manures were looked upon not as wastes but as organic fertilizer. This cycle probably was never completely closed on most farms because livestock products were shipped off the farm and some of the manurial nutrients were lost to the atmosphere.

In an effort to get higher production out of the system, chemical fertilizer, pesticides and energy are inputs to modern farms. Furthermore, inexpensive chemical fertilizers are often viewed not only as supplements to manurial nutrients but as substitutes for them. Figure 3 shows the break in the basic

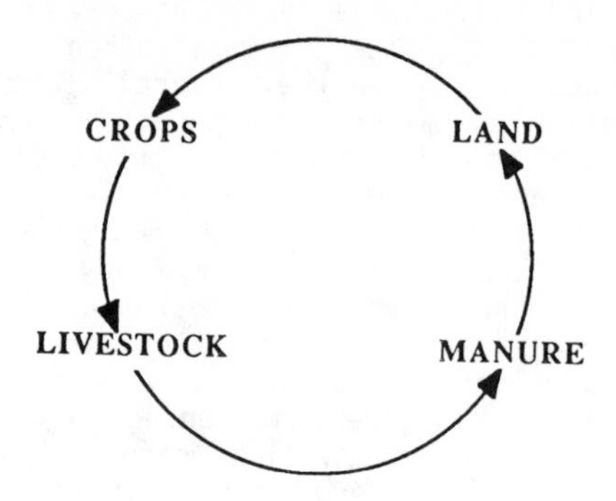

Figure 2. Livestock-crop nutrient cycle.

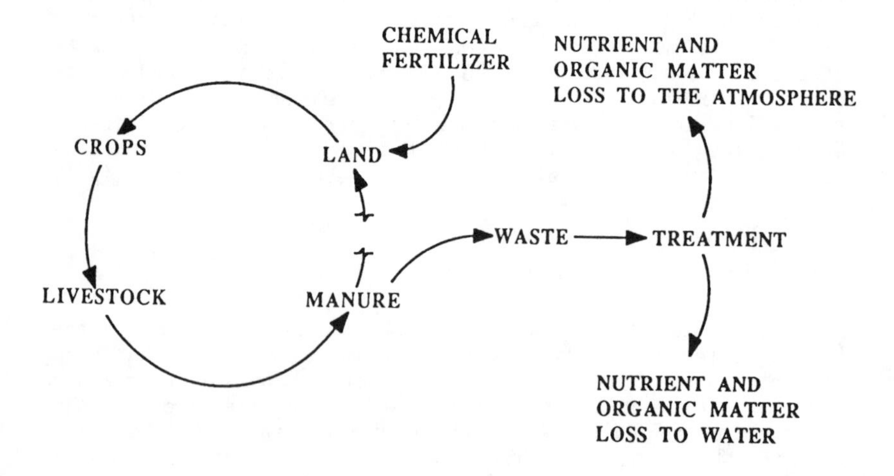

Figure 3. Livestock-cropping system with gap due to relatively inexpensive fertilizers.

livestock-crop nutrient cycle. "Inexpensive" as applied to fertilizer is a term relative to the cost of using manurial fertilizer. Modern monocultural production often concentrates livestock on one site, and imports feed from outside the farm unit. The cost to return this manurial fertilizer to cropland becomes prohibitive if it is hauled very far, and many livestock farmers end up with more manure than their own crops can use.

Many research and educational programs over the last 20 years have been aimed at mending the break in the cycle so that manures will be efficiently used on modern farms. High labor requirements for manure handling, increased travel distance for spreading, and reduced opportunities for use of the manurial resource on the farm unit have all tended to widen the gap. Nonetheless, management systems have been developed that adopt technologies to reduce labor in handling manure and increase its nutrient value so that the cycle on modern livestock farms is not broken completely. In any case, if the cycle is broken the manurial resource becomes a waste problem which often has a least cost-solution within the context of land application.

If indiscriminately used, conservation tillage on fields where manure is spread could drive yet another wedge in the already broken nutrient cycle. For conservation tillage on livestock farms to be an efficient farming practice that will control soil erosion, prevent water pollution, and conserve manurial resources, it must be tailored to the specific livestock farming system.

NUTRIENT MOVEMENT IN MANURE MANAGEMENT SYSTEMS

The objectives of a tillage-manure management system are often different for farmers than they are for environmental and natural resource protection agencies. While the objectives are not always the same for these two groups, there may be systems that are seen as beneficial to both.

Figure 4 is a simplified schematic illustrating four possible paths that manurial nutrients might take after being applied to land. Value judgments from two groups, farmers and environmental protection agencies, about the effects of manurial nutrients moving in each path are identified.

The sketch suggests that both farmers and environmental protection agency personnel would agree that the best possible fate for manurial nutrients is its utilization by crops. This is at least one common objective. The two groups may differ on whether or not manurial nutrient loss to the atmosphere is good or bad. If there is a high probability that some nutrients will be lost to the surface or groundwater, environmentalists probably would prefer it be lost to the atmosphere. On the other hand, farmers would likely prefer that nutrients not be lost to the atmosphere so that a larger pool of nutrients is available to the crops, even if some nutrient is potentially lost to runoff or deep percolation. Both groups would not like to see any manurial nutrients lost to the water, but environmentalists probably would feel strongest on this issue. Figure 4 reveals that the objectives of farmers and environmental agencies have quite a lot in common.

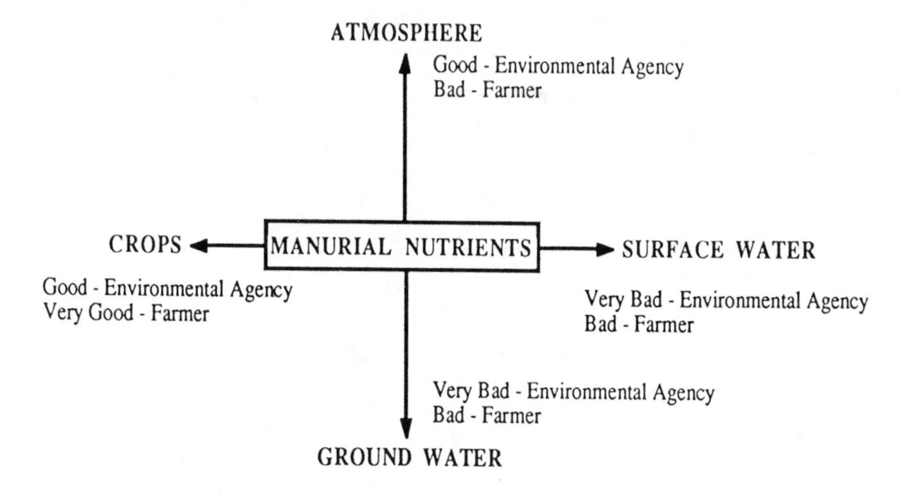

Figure 4. Simplified sketch of manurial nutrient movement after land application.

In assessing the effects of conservation tillage on the simplified system shown in Figure 4, we can consider a long-term manurial nutrient balance such as that given below:

> Manurial nutrients applied = manurial nutrients lost to atmosphere + manurial nutrients lost to surface water + manurial nutrients lost to groundwater + manurial nutrients used by crops

Further we need to consider how conservation tillage might affect the magnitude of nutrient flow in each component of the balance.

First, ammonia losses to the atmosphere from manure would appear to be somewhat greater using conservation tillage rather than conventional moldboard plow. Once surface-spread manure is plowed under there is little ammonia loss. If manure is injected in combination with conventional or conservation tillage, volatile losses would also be expected to be low (Hoff, et al. 1981).

Losses by combining surface-spreading with conservation tillage are less certain, but should be greater than by injection or by immediate plow down after surface application.

If the manure were left on the surface, substantial ammonia volatilization would be expected (Beauchamp, et al. 1982). In fact, the losses from applying the manure to a residue-covered field might be higher than to bare ground (Donovan and Logan, 1983). Several types of conservation tillage, such as disking or chiseling would provide some incorporation.

A study at Cornell University in 1982 (IDES, 1986) found that the chisel plow did almost as well in covering surface applied liquid dairy manure with soil as the moldboard plow (97% as compared to 100%). Whereas the manure may be covered with soil by the chisel plow, the depth of incorporation is much shallower than with the moldboard. This was evidenced in the same study by higher ammonia fluxes after tillage from the chiseled plot than the moldboarded plot. However, considering the substantial reduction in anhydrous ammonia loss by only several centimeters of incorporation (Jackson and Chang, 1947), the light incorporation of manure by disking or chiseling may reduce ammonia losses to 25% or less of the ammonia present in the applied manure.

Any increase in ammonia loss from manure incorporation by chisel plow apparently has little effect on use of manurial nutrients by the crops. Figure 5 shows a summary of three years of corn yield data for a study done in central New York State as part of the Integrated Dairy Energy Systems Research Project (IDES).

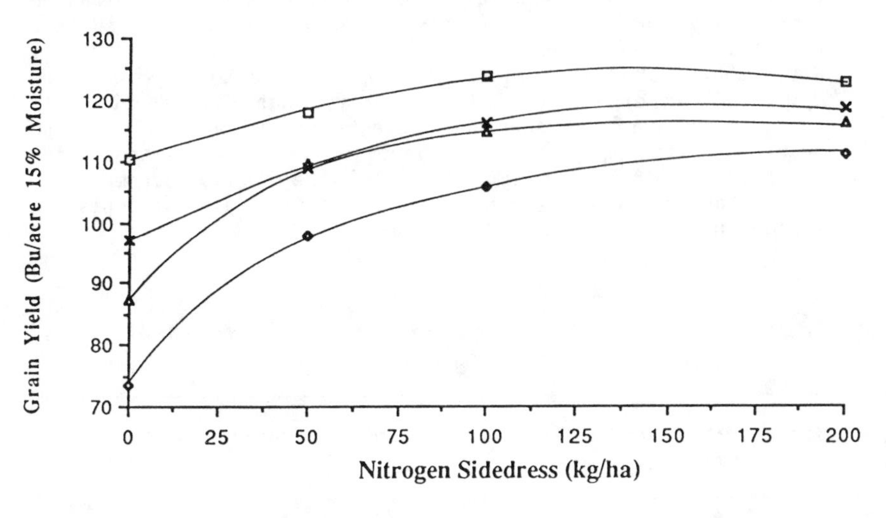

Figure 5. Corn grain yields for different tillage, manure management and sidedress N levels (IDES, 1986).

The figure shows that corn yields for moldboard plowing were higher than for chisel plowing but that is not necessarily central to the topic of this section. It is important to note that the relative yield increase due to the addition of manure at a rate necessary to meet crop nitrogen requirements is almost identical for both tillage systems. These data suggest that the amount of manurial nutrients (and in particular nitrogen) taken up by crops is the same for moldboard and chisel plowing. However, further work at different locations and with a variety of reduced tillage implements are necessary to confirm these results.

More extensive studies have looked at manurial nutrient losses in surface runoff under reduced tillage systems. For the past several years we have been conducting a study as part of the Great Lakes phosphorus control program that has sought to quantify manurial phosphorus losses in surface runoff as affected by various tillage systems. This study has focused on the influence of incorporation depth as well as rate of manure application, time of runoff event relative to application, and interactions among these three.

All forms of manurial nutrient loss in runoff were greatly reduced by incorporating (mixing) the manure to a depth of 3 cm. Further incorporation to a depth of 10 cm resulted in less dramatic but still further reductions. Table 1 shows that the effects of incorporation on TP, TSP losses from subsequent runoff events were much less after the initial precipitation event. Apparently, the nutrients susceptible to loss to surface runoff are either transformed or "fixed" to the soil in some way after the manure has been applied and allowed to dry for a few days. Other experimental tests verify that the reduction in losses from runoff in subsequent events is not primarily due to nutrient loss in the first event. These experiments have verified the relative importance of the first increment of incorporation on preventing losses of manurial nutrients in surface runoff.

Table 1. Effects of Incorporation of Manure on TP/TSP Load (percent of maximum load)

Incorporation Depth (cm)	Precipitation Event				
	1	2	3	4	5
15	4/3	3/2	5/3	1/1	6/4
10	8/7	3/3	4/4	1/1	7/4
3	19/20	10/7	4/4	2/1	3/3
0	100/100	29/52	4/6	2/2	13/6

We found that very heavy applications of manure left unincorporated greatly reduced surface runoff. These results indicated that incorporation by disk or chisel are almost as effective as a moldboard plows at reducing surface runoff losses. Data in Table 2 shows levels of nitrogen and phosphorus (mg nutrient/kg manure applied) found in surface runoff from a portion of the Great Lakes study. These values are averages obtained from laboratory experiments using three surface rates of manure application; 135, 67 and 22 mt/ha. The first runoff event occurred within an hour of manure application. Table 2 shows that, although the concentration of total nitrogen or phosphorus might be higher in runoff from heavy applications of manure, the total loss per unit of manure applied is less. Therefore, it is possible to have very little total nutrient loss in surface runoff even if manure is not incorporated. Since manure is often applied to fields that have been harvested for silage, the manure is sometimes the primary source of surface residue. Table 3 shows a comparison of equivalent surface "residue" from manure applications for several tillage practices. This residue effect, in part, influences runoff potential and nutrient losses. It is important to consider that these heavy application rates mean that a smaller acreage would receive manure each year, so that total surface runoff losses from a farm might well be decreased by this application technique. Lower transportaion, energy, and labor costs could provide benefits to the farmer as well.

Table 2. Total Phosphorus and Total Kjeldahl Nitrogen in Surface Runoff, Normalized (mg/kg manure applied).

Rate (mt/ha)	Initial Runoff Event (mg/kg)		Mean Value of Subsequent Precipitation Events (mg/kg)	
	TP	TN	TP	TKN
135.	32.	970.	6.	45.
67.	71.	520.	10.	48.
22.	188.	710.	54.	246.

Table 3. Equivalent Residue Cover for Selected Tillage Manure Systems

Field Operation	% Residue Cover Remaining After Each Operation[1] (%)	Surface Application of Manure (mt/ha) to Achieve Equivalent Residue Cover[2]	
		Without Bedding	With Bedding
Moldboard	3–5	25	7
Chisel			
Straight	50–77	230	60
Twisted	30–60	135	36
Disk			
7.5 cm Deep	40–70	210	56
15.2 cm	30–60	135	36
Field Cultivator	50–80	352	93
Planters			
No coulter	90–95	660	174
Narrow Ripple coulter	80–85	80	128
Double Disk Furrowers	60–80	350	93

1. Conservation Tillage Information Center, 1985.

2. Lake Ontario Phosphate Evaluation Program, 1986 and Robillard and Walter, 1986.

Finally, we know very little about the influence of tillage on manurial nutrient loss to groundwater and have found no direct measurements comparing conservation and conventional tillage and their relative effects on manurial nutrient loss to groundwater. However, a mass balance approach can be used to estimate groundwater losses if the other values are known.

Gilbertson, et al. (1979) estimated nitrate leaching losses from land applied manure using a modeling technique applied to various land resource areas. Relatively high leaching losses of nitrate when manure is applied at rates greater than recommended for crop uptake are predicted. For example, if four times the recommended manure rate is used, the predicted nitrate leaching loss increases by nine times in terms of "pounds N per 100 pounds of crop content." We expect, however, that the actual amount of nitrogen taken up by the crop would be considerably more than that needed for maximum yield.

Therefore, the nitrogen leached might be less than predicted. This and similar models used for estimates of leaching losses might have a signifcant error. Most currently used models assume that the soil is homogeneous. This assumption approaches the actual field situation when the fields are tilled and the macropores are periodically destroyed. However, for no-till the homogeneous soil assumption is not valid. In an experiment that we performed in a no-till field, 5 cm of water moved the chemicals at the surface to a depth of 2 cm in some cases. Homogeneous flow models, such as that of Gilbertson et al. (1979) would have predicted a depth approximately 20 cm. Thus, these models should be viewed with some caution. The effect of conservation tillage on manurial nitrate leaching depends on its direct effect on water penetration below the root zone. Woolhiser (1975) suggests that conservation tillage will reduce surface runoff slightly to moderately which will result in a somewhat smaller increase in deep percolation.

MANAGEMENT ALTERNATIVES

We have seen that tillage practices can greatly affect nutrient losses to surface water, and may to some extent affect nitrogen losses to air and groundwater. The experience suggests that strategies could be developed to mitigate the negative impact of those losses in a reduced tillage system. Where the nitrogen benefits of manure are particularly important to the farmer, chisel plow incorporation offers an option to moldboard plowing with similar crop yield benefits. Where manure is in abundant supply and nitrogen conservation less critical, heavy but infrequent surface applications of manure might be an attractive labor and energy saving option offering possible beneficial control of surface runoff losses.

The practice of conservation tillage has so many positive attributes and every effort should be made to try to effectively integrate them into livestock farming systems. For example, if manure can be applied at high rates to land once in a three to five year rotation cycle it might be possible to develop a hybrid tillage system that used a moldboard plow to incorporate manure the year it is spread and conservation tillage during the other years. This system would appear particularly appropriate where weed and insect problems have accumulated through the conservation tillage period. It also would incorporate manurial phosphorus deeper into the soil profile. Phosphorus will tend to accumulate near the surface of the soil if reduced tillage is used and manure application is made at rates based on crop nitrogen needs.

Conservation tillage should work very well with manure injection systems. Injection of semi-solid and liquid wastes is an extremely effective technology for nutrient conservation on the farm. Because the wastes are injected below the surface, surface runoff and atmospheric losses of nutrients are minimized. The specialized equipment required and energy inputs are expensive and custom applicators can charge rates equivalent to the fertilizer value of the nutrients, so that the economic incentives for nutrient recycling may not be as attractive as other systems. Manure injection might be expected to increase groundwater pollution in some cases. Also, some detrimental effects on crop production have been noted (Hoeft and Vanderholms, 1983).

NUTRIENT CONSERVATION: HOW MUCH IS ENOUGH?

Each of the application technologies outlined above can potentially be managed in a manner which protects water quality for downstream users. The question of which strategy farmers choose will thus largely depend on their own balance of the benefits of maximal nutrient availability to the crop versus the costs of application at that optimal rate. In general, we do not wish to challenge time-proven notions about "good" manure management, but we feel that the data indicate that in some situations manurial nutrient conservation may be given too much emphasis. Gilbertson, et al. (1979) state that, "If recommended rates of manure are not used, economic losses incurred through loss of nutrients will become more significant as fertilizer costs increase." While this might be true generally, we have found that many, if not most, of New York State dairy farms cannot efficiently use all of the manurial nitrogen, let alone phosphorus produced on the farm. The United States census of dairy farms suggests that in general more manurial nitrogen is available on the average dairy farm than can be used on that farm. Manurial phosphorus is greatly in excess of what can be used. Many farmers will not be able to follow recommendations for manure application rates based on crop phosphorus requirements such as those currently proposed in Wisconsin (Massie and Peterson, personal communication). Expensive transportation costs often reduce cost-effective use of manure off the farm where it is produced, or even on the farm if the land area is large or spread out.

The Livestock Waste Facilities Handbook (1985) emphasizes the need to avoid applying manure in excess of crop nutrient needs. For example, it states that, "Applying excess waste can harm crop growth, contaminate soil, cause surface and groundwater pollution and waste nutrients," (p. 101). In many cases such concerns are justified. However, in some situations they may not be. For example, as mentioned above, the waste of nutrients depends on whether or not there is a crop that can economically use them. Crop yields can be reduced as a result of excess application of manure. However, application rates considerably higher than those based on nitrogen requirements might not result in decreased yields. Gilbertson, et al. (1979) shows that corn forage yields increased (up to 135 mt/ha (wet-weight basis)) as a result of annual manure application. Soil contamination from trace metals or salts in animal waste can result from frequent applications over a long period of time. However, heavy applications of manure need not contaminate the soil, particularly if they are not repeated each year. If bedding is added, lack of available nitrogen to crops due to a high C/N ratio shortly after application could be a major concern in some cases. This would be partly offset by the starter sidedressed nitrogen fertilizer that is generally recommended whenever the manurial nitrogen is used.

The concern that manure applications in excess of crop requirements will cause water pollution needs to be considered on a case by case basis. In general, the heavier the rate of application to an area, the greater the potential for nutrients to be lost from surface runoff or groundwater movement from that area. However, it is not nearly as obvious that if the manure is spread over a larger area, less will be delivered to water. In fact, our data suggest that in some cases more phosphorus and nitrogen will be lost to surface water if the same quantity of manure is applied over a larger area. For example, in laboratory and field studies, the concentration of nutrients in runoff increases as the rate of application increases, but at a much lower rate. Furthermore, once the surface applied manure dried, total surface runoff decreased for higher manure applications.

Groundwater contamination, particularly from manurial nitrate, might still result from heavy manure application. We have no good comparative data on this. In general, if surface runoff is decreased, deep percolation will increase but by a smaller amount. One would expect a similar situation to be true for nitrogen. That is, if less nitrogen is lost in surface runoff, more might be expected to percolate to the groundwater but more would be immobilized by plants to denitrify.

Consideration of the many options available for manure application/disposal are important for determining the effectiveness of conservation tillage on dairy farms. Admittedly, data on manure conservation tillage systems are limited, but enough are available to suggest that under some circumstances conservation tillage might work well with the practice of surface application of manure. Water quality issues seem to be of great concern when liquid waste is surface applied and immediately followed by a runoff event. Moldboard plowing after surface application of manure, but before the runoff event, greatly reduces nutrient loss in surface runoff. Chisel plowing is less effective in reducing the overland nutrient loss, in any case, and is relatively ineffective as a tillage practice while the surface is covered with liquid manure. If manure is handled in a solid form or in a way that allows the ammonia to be lost to the atmosphere, the potential for either surface or groundwater contamination seems no greater with conservation tillage than with conventional tillage. Furthermore, the effectiveness of the manurial nutrients for crop production seems similar, runoff is reduced and in general the soil erosion control of this practice is enhanced.

SUMMARY AND CONCLUSIONS

Our experience with tillage-manure systems leads us to the following six summary statements:

1. Conservation tillage and manure management are compatible.
2. Effects of manurial nitrogen on crop yields are the same for both moldboard and chisel plowing.
3. Losses of manurial nutrients in surface runoff are greatly reduced by a little incorporation.
4. Manurial residue can significantly reduce runoff from cropland where residue has been removed.
5. Management costs of utilizing manurial nutrients as perceived by the farmer can be extremely high.
6. Because manure is not generally utilized as a fertilizer source, there is a potential groundwater problem with any tillage system.

As farmers and environmental regulators begin to look at manure management with conservation tillage, it will be important to look closely at the total farming system. Resources available on the farm must be balanced with outside inputs, and livestock waste must be viewed as both a waste management challenge and a crop nutrient resource. It will be crucial to examine systems as they operate, in a particular climate and with a particular combination of livestock and cropping system. When all these concerns are addressed, it should be possible to develop a strategy which both efficiently allocates farm resources and protects downstream users.

While farmers want to maximize the nutrient pool available to plants, environmental regulators often try to minimize this nutrient pool to provide less risk of runoff or groundwater pollution. Prescriptive application rates from either perspective thus far have assumed a steady state situation, but in fact nutrient fate is dependent on a variety of stochastic processes, including climatic uncertainties, variation of crop-type and growth pattern, and their interaction with the soil system. Nonetheless, research results indicate that when manure is surface applied and allowed to dry before a precipitation event, much higher than usual application rates can be tolerated without harmful effects on water quality. Somewhat higher losses per hectare are offset by a smaller area over which the manure is applied, with net benefits to all concerned. When an abundance of manure lessens the need for maximum nutrient conservation, high application rates with no incorporation may not lead to the surface or groundwater contamination commonly feared. In any case, systems using reduced tillage are compatible with good manure management. Efficient farm operation and water quality protection can both be achieved with careful management.

REFERENCES

Beauchamp, E.G., G.E. Kidd, and G. Thurtell. 1982. Ammonia volatilization from liquid dairy cattle manure in the field. Can. J. Soil Sci. 62:11-19.

Cornell Recommends for Field Crops. 1985. Department of Agronomy, Cornell University, Ithaca, NY.

Conservation Tillage Information Center. March 1985. League City, TX.

Donovan, W.C. and T.J. Logan. 1983. Factors affecting ammonia volatilization from sewage sludge applied to soil in a laboratory study. J. Envir. Qual. 12:584-590.

Gilbertson, C.B., F.A. Norstadt, A.C. Mathers, R.F. Holt, A.P. Barnett, T.M. McCalla, C.A. Orstad, and R.A. Young. 1979. Animal Waste Utilization on Cropland and Pastureland. EPA-600/2-70-59.

Gunkel, W.W. 1974. Energy inputs for agriculture. Department of Agricultural Engineering. Cornell University. Ithaca, NY.

Hoeft, R.T. and D.H. Vanderholm. 1983. Evaluation of detrimental effects on crop production from manure application. ASAE Paper #83-2122.

Hoff, J.D., D.W. Nelson, and A.L. Sutton. 1981. Ammonia volatilization from liquid swine manure applied to cropland. J.E.Q. 10:90-95.

Integrated Dairy Energy System Project, Final Report-1986. Department of Agricultural Engineering, Cornell University, Ithaca, NY.

Jackson, M.L. and S.C. Chang. 1947. Anhydrous ammonia retention by soils as influenced by depth of application, soil texture, moisture content, pH value, and tilth. Agron. J. 39:623-633.

Livestock Waste Facilities Handbook. 2nd Ed. 1985. MWPS-18. Iowa State University, Ames, IA.

Lake Ontario Phosphorus Evaluation Program. 1986. New York Department of Environmental Conservation. Albany, NY.

Robillard, P.D. and M.F. Walter. 1986. The influence of tillage on phosphorus losses from manured cropland. Final Report to the New York State Department of Enviornmental Conservation, United States Environmental Protection Agency, Department of Agricultural Engineering, Cornell University, Ithaca, NY.

Woolhiser, D.A. 1975. Control of water pollution from cropland. Vol. 1, Chapter 3. EPA-600-2-75-026a.

CHAPTER 16

AN ASSESSMENT OF GREAT LAKES TILLAGE PRACTICES AND THEIR POTENTIAL IMPACT ON WATER QUALITY

T. J. Logan
The Ohio State University, Columbus, Ohio

INTRODUCTION

A major objective of the workshop was to discuss the impacts on water quality of accelerated conservation tillage adoption in the Greak Lakes Basin — specifically, whether there are adverse effects on surface and groundwater contamination by nitrate and pesticides.

Eutrophication of the Great Lakes has been a major environmental issue since the mid 1960's when Lake Erie was declared to have "died". It has been generally accepted that the cause of accelerated phytoplankton growth was phosphorus loadings. Long-term studies in the 1970's (PLUARG, 1978; COE, 1982) showed that phosphorus loadings to the lakes were primarily attributable to wastewater treatment plant discharges and to runoff from land, the specific percentages varying from lake to lake. Most of the phosphorus in land runoff was found to be associated with sediments eroded from agricultural land devoted to intensively cultivated annual crops with lesser amounts from livestock waste runoff. Analysis of remedial measures available for reduction of cropland runoff containing phosphorus indicated that significant decreases could be potentially achieved through wide-scale adoption of conservation tillage practices by Greak Lakes Basin farmers (Forster et al., 1985). The rationale for this strategy is the high effectiveness of conservation tillage in reducing sediment loads from agricultural land and the concomitant reduction of particulate phosphorus which often represents 80% or more of the total phosphorus load. This approach was initially proposed by

Effects of Conservation Tillage on Groundwater Quality: Nitrates and Pesticides, Terry J. Logan et al., eds.

the Pollution From Land Use Activities Reference Group (PLUARG) to the International Joint Commission (IJC) for the Great Lakes (PLUARG, 1978) and by the U.S. Army Corps of Engineers Lake Erie Wastewater Management Study (1982). Subsequent to these recommendations, the individual Great Lakes states and provinces were charged with developing plans for reducing their share of the target phosphorus loads to specified limits (e.g., Ohio EPA, 1985). Most of these plans, and in particular those of Ohio and Michigan, rely on accelerated implementation of conservation tillage as a means of meeting a major part of the target load reductions. In Ohio, for example, this will require about 470,000 hectares (1,000,000 acres) in conservation tillage.

As discussed in the workshop, there is some public sentiment that we may be substituting one environemntal problem for another -- increasing nitrate and pesticide contamination of surface and groundwater in order to prevent agricultural phosphorus losses and to control erosion. While the data in favor or against this view have been reviewed in detail in the previous chapters, we present here the specific perspective of the Great Lakes region, i.e., the impacts of climate, soils, crops, specific cultural practices, and weed and insect pressures on tillage practices and nitrogen fertilizer and pesticide use; and the effects of these variables on the potential for nitrate and pesticide contamination of surface and groundwater in this region.

THE GREAT LAKES BASIN -- SPECIFIC INTERACTIONS OF TILLAGE AND WATER QUALITY

Climate

Most of the Great Lakes region is dominated by continental and lake-modified continental climates with rainfall exceeding evapotranspiration in the late fall through late spring period. Runoff is greatest in the early spring during and after snow melt when soils are generally saturated, with lesser amounts occurring in late spring and in late fall. Percolation as indicated by tile drainage (Logan, 1981) is greatest in the early spring after ground thaw and in the late fall with the onset of frontal storms. Nitrogen fertilizer and pesticide use is concentrated in the period March-June in conjunction with corn and soybean planting. This coincides with the period of maximum precipitation, runoff, and tile flow. Water movement to deeper groundwater would also be expected to be high during this period, although there have been no deep groundwater studies in this area.

The potential of no-till, and to a lesser extent conservation tillage, to increase infiltration and percolation as discussed in previous chapters is less on the poorly-drained till soils of much of the Great Lakes Basin than on coarser-textured soils because of the lower saturated hydraulic conductivities of these soils (Logan and Adams, 1981; Logan,

1981). In the case of nitrogen, high denitrification potentials of these poorly-drained soils will decrease the possibility of nitrate leaching, especially in the spring when soils are wettest. Of greater importance to nitrate leaching might be late fall and early spring movement of residual nitrate out of the root zone, but this process will be more related to rates of nitrogen fertilizer use on corn rather than on tillage type.

Temperatures above 4°C in the spring when critical biological processes such as denitrification can occur are consistently reported from late March in the southern parts of the Basin to mid-April in more northern regions. In the fall, biological activity is minimal by late October in northern areas and by late November in the southern part of the Basin.

Soils

Great Lakes Basin soils were predominately formed from glacial deposits of recent origin (Wisconsin as glacial tills are often < 15,000 years old), and are generally fine-textured, poorly-drained, and fertile. Exceptions to this are the coarse-textured sands of Michigan and the non-glaciated areas of the eastern Lake Erie Basin and the Lake Ontario Basin. Specific consequences of note from these general soil properties are intense use for grain crop production, extensive use of subsurface drainage, and high background losses of nutrients in surface and subsurface drainage, particularly losses of phosphorus.

The poor drainage of many of the soils of this region and the high net annual percolation make the Great Lakes Basin unique for extensive use of subsurface tile for drainage. The consequences for chemical transport are that constituents such as nitrate and soluble pesticides will tend to be transported with tile flow rather than in surface runoff (Logan et al., 1980; Baker and Johnson, 1976). As tile drainage is discharged to surface ditches, it becomes part of stream flow and can, therefore, contribute to surface water contamination. At the same time, interception of subsurface flows by tile lines will prevent some soluble chemical movement to deep groundwater.

Crops

The Great Lakes Basin constitutes a major part of the grain-producing agricultural sector of the U.S. where major crops are corn, soybeans, and wheat. It is also a major area for crop production in Ontario and for specialty crops in Michigan. Livestock-based farming is concentrated in Wisconsin, Minnesota, and New York. In terms of chemical use, almost all of the nitrogen fertilizer is applied to corn in the Basin; the Great Lakes states produce 37% of U.S. corn and account for 21% of U.S. consumption of nitrogen fertilizer. If there is a trend towards more rotation of the three principal grain crops, corn, soybeans, and wheat, as suggested in the workshop, the potential for nitrate leaching should decrease. Logan et al. (1980) and Logan (1981) have shown that the higest nitrate concentrations and loads in tile drainage occur with corn and are much lower with soybeans or wheat. The higher nitrate concentrations with corn were attributed directly to rate of nitrogen fertilizer application in excess of crop utilization (Logan, et al., 1980). Rate, timing, and placement of nitrogen fertilizer will have to be made more precise to avoid increasing nitrate leaching potential as nitrogen rates increase. In the last decade or so, nitrogen fertilizer use for corn in Great Lakes Basin states has increased to a present average of 155 kg N/ha.

Cultural Practices

The three principal crops grown in the region, corn, soybeans, and winter wheat, are often rotated, although some continuous corn is also grown. About 30-50% of the corn and soybean acreages in the Basin are in a corn/soybean rotation. Fall moldboard plowing or fall chisel plowing are commonly used, particularly in the glacial till areas where the soils tend to be more poorly-drained and where spring tillage is consequently difficult. No-till and conservation tillage have been shown (Van Doren et al., 1976) to produce higher crop yields where soils have better drainage or where drainage can be improved by tile installation. Several of the Lake states, including Indiana and Michigan, and the province of Ontario do not recommend no-till on the more poorly-drained soils. Current statistics (see Chapter 2) indicate that about 50% of the U.S. Great Lakes cropland acreage is in some form of conservation tillage with most of that being mulch tillage; only 20% of conservation tillage acreage is no-till and much if that is cultivated during the growing season. There is little continuous no-till, but no-till corn/soybeans rotation is likely to be an increasingly popular combination.

A recent innvoation in the Great Lakes Basin -- but one that is likely to gain in popularity -- is ridge tillage where single row ridges are intitially formed after plowing -- usually in the fall -- and are reshaped each year for several years afterwards by row cultivation after crop emergence. The advantage of this system is perceived to be increased surface drainage and warmer soil temperatures in the vicinity of the ridge during the critical period of germination and seedling growth and the elimination of fall tillage subsequent to ridge formation. The need to cultivate ridges to maintain shape should reduce the dependence on herbicides for this system compared to no-till. Development and use of equipment to band herbicides in the ridge will reduce the potential for herbicide loss by volatilization and runoff.

Weed and Insect Pressures

The workshop attendees were of the opinion that weed and insect pressures were generally lower in the northern grain-producing states than in the south. As a result, pesticide use is not expected to increase significantly with adoption of conservation tillage, with the possible exception of "knock-down" herbicides to control early germinating weeds controlled by tillage in the conventional tillage systems. Also, insect damage to the grain crops grown in the month is not as severe as it is on other crops such as cotton, and the statistics show that insecticide use is small compared to herbicide applications. Rotation of the three major grain crops probably aids in reducing both weed and insect pressures.

Summary

Water quality problems associated with agricultural production are of particular concern in the Great Lakes Basin because of the intensive use of lands in this area for grain crops, the high net precipitation over evapotranspiation which results in significant runoff and leaching, and the proximity of these lands to one of the world's largest bodies of freshwater.

Conservation tillage is being promoted in the Basin for control of nonpoint sources of phosphorus to the Great Lakes and the data indicate that Basin farmers are adopting this practice at a steady rate although most are shifting to a system minimizing tillage rather than to strictly no-till.

Corn, soybeans, and winter wheat are the major crops of the region and, of these, only corn receives nitrogen fertilizer at rates which could contribute significantly to losses of nitrate in runoff and tile drainage or movement to groundwater. This potential for nitrate contamination of surface and groundwater with corn fertilization is a consequence of less-than-precise nitrogen fertilizer recommendations and less-than-efficient timing of fertilizer applications, regardless of the tillage system used. Accelerated adoption of conservation tillage is, therefore, unlikely to have much of an impact on these losses except to the extent that changes in hydrology with tillage may change the relative losses of nitrate to surface versus groundwater. The increased infiltration, and potential for increased groundwater contamination, observed with no-till on well-drained soils is less likely to occur on the more poorly-drained till soils of the Great Lakes Basin where runoff and tile drainage are greater than movement to groundwater.

Weed and pest pressures on the major grain crops of the Basin are relatively low compared to conditions in the southern U.S.. Rotation of the three major grain crops also serves to reduce build up of weeds, diseases, and insects. CTIC estimates of pesticide use by Lake Erie Basin farmers (see Chapter 3) indicated that there was very little change in quantity and type of pesticide used with various tillage systems except for some increased use of "knock-down" herbicides with conservation tillage. With the low overall use of no-till in the Basin compared to other types of conservation tillage, the impact of accelerated conservation tillage use by Basin farmers on pesticide contamination of water supplies is likely to be small with the greatest effect being seen in surface rather than groundwater.

REFERENCES

Baker, J.L., and H.P. Johnson. 1976. Impact of subsurface drainage on water quality. Proc. Third National Drainage Symp. Am. Soc. Agric. Eng., St. Joseph, MI.

Corps of Engineers. 1982. Lake Erie Wastewater Management Study. Final Report. Buffalo District, Buffalo, NY. 22 p.

Forster, D.L., T.J. Logan, S.M. Yaksich, and J.R. Adams. 1985. An accelerated implementation program for reducing the diffuse source phosphorus load to Lake Erie. J. Soil Water Conserv. 40:136-141.

Logan, T.J., G.W. Randall, and D.R. Timmons. 1980. Nutrient content of tile drainage from cropland in the North Central Region. North Central Regional Research Pub. No. 268. OARDC, Wooster, OH. 16 p.

Logan, T.J. 1981. Maumee River pilot watershed study. Continued watershed monitoring (1978-1980). Great Lakes National Program Office. USEPA. EPA-905/9-79-005-C. 56 p.

Logan, T.J., and J.R. Adams. 1981. The effects of reduced tillage on phosphate transport from agricultural land. Lake Erie Wastewater Management Study Technical Report Series. U.S. Army Corps of Engineers. Buffalo District, Buffalo, NY.

Ohio Environmental Protection Agency. 1985. State of Ohio phosphorus reduction strategy for Lake Erie. Ohio EPA, Columbus, OH. 89 p.

Pollution From Land Use Activities Reference Group. 1978. Environmental management strategy for the Great Lakes system. Final Report to the International Joint Commission. Windsor, Ontario. 115 p.

Van Doren, D.M., Jr., G.B. Triplett, Jr., and J.E. Henry. 1976. Influence of long term tillage, crop rotation and soil type combinations on corn yield. Soil Sci. Soc. Am. J. 40:100-105.

SUMMARY AND CONCLUSIONS

J. L. Baker,
Iowa State University, Ames, Iowa

T. J. Logan,
The Ohio State University, Columbus, Ohio

J. M. Davidson,
University of Florida, Gainesville, Florida

M. Overcash,
North Carolina State University, Raleigh, North Carolina

The purpose of the this book is to evaluate positive and negative impacts of conservation tillage systems on the quality of surface water and groundwater with respect to pesticides and nitrate, and, where negative impacts may exist, to identify corrective practices.

Increased adoption of conservation tillage is predicted: Conservation tillage use is expected to increase past the year 2000, as the benefits of soil conservation as well as savings of time and energy are realized. The most significant feature of conservation tillage is the residue left on the soil surface from the previous crop: one accepted definition of conservation tillage requires 30% or greater residue coverage at the time of planting. This level may be hard to achieve for some crop-tillage-harvest-climate conditions, but conservation tillage advantages may still exist at somewhat lower residue coverages.

Effects of Conservation Tillage on Groundwater Quality: Nitrates and Pesticides, Terry J. Logan et al., eds.

Strict no-till, an extreme form of conservation tillage, is less likely to be used on a continuous basis: Tillage systems range from "conventional" moldboard plow-disk-plant systems, where essentially all crop residue is buried in disturbed soil, to the strictly undisturbed no-till system. It was considered that any mechanical soil disturbance, such as chisel-plowing, disking, or other forms of cultivation, results in an intermediate tillage system, with soil conditions more closely related to the conventional system (i.e., the no-till system is an extreme case, particularly if used for several successive years, so that new soil conditions can be established). A review of no-till production practices currently in use revealed that mechanical cultivation is often performed for weed control and that tillage systems, like crops, are sometimes rotated so that a particular piece of ground will not likely be in continuous no-till. Therefore, strict no-till conditions, while of research interest, may not exist in significant quantity in the farming community in the near future.

Pesticide use in the Corn Belt should not increase significantly with increased conservation tillage use: It is frequently stated that the advent of herbicides has allowed conservation tillage to work; however, it is clear that weed and insect control is important for all tillage systems. This struggle against pests requires the use of all available tools, and current surveys show that herbicides have not replaced but instead supplement mechanical weed control, even in conventional tillage systems. Therefore, rates of pesticides already in use with conventional tillage are not expected to change as a farmer converts to conservation tillage, except possibly in the short term, where an uncertainty in pesticide requirements may result in the use of larger quantities. The one additional pesticide that may be needed is a "knock-down" herbicide to replace primary tillage, to control weeds that emerge before planting. However, erosion control provided by conservation tillage should limit runoff losses of the strongly-adsorbed herbicides currently available, namely glyphosate and paraquat.

Because of less soil disturbance, there are fewer weed seeds in the lower soil profile and more weed seeds in the upper soil profile with conservation tillage; therefore, the weed ecology may change as conservation tillage is adopted. This change may not be bad, but one that should be watched.

Reduced chemical incorporation into the soil with conservation tillage can be a disadvantage: One disadvantage of conservation tillage is reduced potential for the use of soil-incorporated pesticides.

Some pesticides require incorporation in order to reduce or prevent volatilization or photodegradation. Decreased incorporation of pesticides (and nutrients) results in the presence of higher concentrations of the chemical in the soil surface "mixing zone," where the chemical may be released to runoff; this release in turn results in higher chemical concentrations in the runoff. Surface-applied pesticides intercepted by crop residue on the soil surface may interact less with that crop residue than with soil, and therefore are more susceptible to runoff and volatilization losses than if they were adsorbed to soil. For these reasons, development of application methods that allow for incorporation of pesticides (and nutrients) without incorporating crop residue should be encouraged.

Nitrogen management (rate, timing, placement, etc.) is more crucial to nitrate leaching than choice of tillage systems: Correct choices of rate, timing, and placement of fertilizer applications are important for improving use efficiency and reducing nitrate leaching, regardless of tillage method used. Except for acid soils, surface applications of urea and urea-based fertilizers are not practiced because of volatilization losses of ammonia. This can be a particular problem for conservation tillage, where incorporation of these fertilizers is less feasible. Placement of nitrogen fertilizer in the vicinity of undecayed crop residue can also result in crop nitrogen deficiencies because of temporary nitrogen immobilization resulting from the decomposition of nitrogen poor crop residues. Development and use of fertilizer incorporation methods should be encouraged, to avoid the need to make up for losses and "tie-up" by applying more nitrogen fertilizer than required by the crop.

More information is needed to potentially take advantage of differences in nutrient cycling in the surface soil with conservation tillage: There is good evidence that nutrient cycling in the surface soil with conservation tillage is different from that with conventional tillage. There appears to be less organic nitrogen mineralization, more immobilization, and more denitrification with conservation tillage systems. This area requires more study, because conservation tillage systems could reduce the presence of nitrates in the plant root zone compared to conventional tillage. In this, as well as other areas of study, it should be emphasized that the differences between tillage systems will be mainly limited to the maximum tillage depth, usually less than 20 cm.

Conservation tillage reduces losses of (strongly adsorbed) chemicals transported mainly with sediment, but probably not of (moderately to weakly adsorbed) chemicals transported mainly in runoff water: The benefits of conservation tillage in reducing soil erosion are well documented: they involve reduction not only of sediment loads themselves, but also of the presence of sediment-associated nutrients and pesticides. Since sedimentassociated nitrogen, primarily in the form of organic nitrogen, frequently dominates total nitrogen losses with surface runoff, erosion control can reduce total nitrogen losses substantially. The same can be said for strongly adsorbed pesticides (and total phosphorus). However, a large number of pesticides in current use are only moderately adsorbed (e.g., alachlor, atraxine, cyanaxine, and furadan), and are transported primarily in solution with runoff water.

Possible reductions in runoff volumes with conservation tillage are highly variable and usually not a major factor affecting surface runoff losses of chemicals: The reduction in volume of runoff water that is generally associated with conservation tillage has the potential to reduce dissolved chemical losses in runoff. However, the volume reduction on an annual basis may be only 20% or less, and may occur more for storms later in the growing season than for the first storm after tillage and chemical application. As a result, chemical loss reduction may not be significant, especially if lack of chemical incorporation increases concentrations in runoff water.

Potential increases in water movement through the soil profile because of increased infiltration could enhance leaching, but other factors can be more important: The increased infiltration sometimes reported for conservation tillage results from soil surface protection against sealing because of the crop residue, the small dams formed by the residue that lengthen the residence time for infiltration to occur, and possibly the existence of more macropores at the soil surface with conservation tillage (particularly for no-till). The potential for increased water movement through the profile with conservation tillage has resulted in concern; chemical management of timing, rate, and placement have the potential for much greater impact on both leaching and surface runoff losses.

Existence of macropores, and movement of water through them, do not necessarily mean more chemical leaching: The increased presence of macropores with conservation tillage has resulted in expressed concern for greater chemical leaching. It should be understood, however, that preferential flow through macropores may be good in some cases and bad in others. If water ponding at the surface and flowing down macropores bypasses chemicals present within aggregates, the total leaching of the chemical will be decreased. If that same water dissolves chemicals at the soil surface and then flows down through macropores, chemical leaching will be increased. The effects of conservation tillage on the volume and route taken by water moving through the root zone, and associated chemical leaching, deserve additional study.

Conclusion

It would seem that the voiced negative aspects of using conservation tillage--potentially greater use of pesticides and nitrogen and greater leaching of chemicals because of increased infiltration and macropores-are unfounded; decreased chemical losses because of reduced sediment losses are well established for strongly adsorbed chemicals. However, a potential water quality problem is posed by losses of the soluble, moderately adsorbed to nonadsorbed chemicals (most modern pesticides, nitrate, and possible ammonium) in both surface and subsurface drainage. For these chemicals, surface application without incorporation presents increased potential for losses in surface runoff. Therefore, application methods that allow incorporation of the chemical without crop residue incorporation are needed for use with conservation tillage. In addition, research is needed to determine management practices to reduce the leaching potential for both conservation and conventional tillage systems, including movement of chemicals from the rooting zone to the aquifer below.

INDEX